Textbook of Forensic Pharmacy

Second Edition

C.K. Kokate
Vice-Chancellor,
KLE University, Belgaum 590 010, Karnataka,
Former Vice-Chancellor, Kakatiya University, Warangal, A.P.
Former President, Pharmacy Council of India.
Former E.C. Member & Chairman,
All India Board of Pharmacy, A.I.C.T.E.

S.B. Gokhale
Former Principal,
SVKM's College of Diploma in Pharmacy,
Vile-Parle (w), Mumbai-400 056

Pharma Med PRESS
An Imprint of
BSP Books Pvt. Ltd.

PharmaMed Press
An imprint of BSP Books Pvt. Ltd.
4-4-309/316, Giriraj Lane,
Sultan Bazar, Hyderabad - 500 095.

Textbook of Forensic Pharmacy
by **C. K. Kokate & S. B. Gokhale**

Tenth Reprint 2023

Published by :

PharmaMed Press
An imprint of BSP Books Pvt. Ltd.
4-4-309/316, Giriraj Lane, Sultan Bazar, Hyderabad - 500 095.
Phone: 040-23445688; Fax: 91+40-23445611
E-mail: info@pharmamedpress.com
www.pharmamedpress.com/pharmamedpress.net

Printed at

Repro India Limited.

Price: Rs. 350.00

ISBN: 978-93-81075-99-9 (Paperback)

Textbook of Forensic Pharmacy

Second Edition

To my wife

> ***Mrs. Kanan*** *- a pharmacist by profession - who has always encouraged me to aim at academic and professional excellence.*

Dr. Chandrakant Kokate

Preface

As a teacher of Forensic Pharmacy for more than three decades at the Bombay College of Pharmacy, Mumbai and University College of Pharmaceutical Sciences, Kakatiya University, Warangal, I have developed passion for learning more about the Regulations, Associations and their activities, as well as, ethical boundaries of our noble and sacred profession. This profession gives us a one-time opportunity in our lives to serve the cause of humanity by making our humble contributions to the health care system of the country.

During my tenures as President, Pharmacy Council of India; Member, Drugs Technical Advisory Board; Executive Committee Member, All India Council for Technical Education; and Chairman, All India Board of Pharmacy, AICTE, I was associated with most of the statutory bodies governing and regulating the profession of pharmacy in the country. I had good opportunity of interacting with different segments of our profession and policy-makers of the country. I had an access to the write-ups on professional meetings, deliberations and amendments to Acts for several years. It is with this strength of confidence backed by my teaching experience of several years, I have ventured to write this book on Forensic Pharmacy with my colleague and trusted friend Prof. S.B. Gokhale for the benefit of students of diploma and degree classes in pharmacy. The book is written in simple and lucid manner with understandable interpretation of various Acts and Rules, without diluting overall impact of the subject.

The text of Forensic Pharmacy is ever-changing and our efforts were to add commas to the text of the subject, thus facilitating free flow of the knowledge of legal principles regulating our profession.

The historical milestones of various pharma-events and salient features of different Acts and Rules pertaining to profession are integral component of the text. The students of pharmacy, we are sure, will find contents of the book interesting and educative.

We thank management of PharmaMed Press, especially, Mr. Anil Shah for his efforts in publishing this book.

C.K. Kokate

Ably assisted and Co-authored by
S.B. Gokhale

Preface to Second Edition

After several reprints to First Edition, this Second Edition of our book is being brought out covering added features of pharmaceutical ethics, legislation and profession. This revised edition has eleven more chapter included along with updated information on events of today and Annexures. The text of the book is student-centric. It is explained in lucid manner, which is understandable by an average under-graduate student of pharmacy and allied sciences.

We wish to thank Prof. Ganachari, Dr. Pramod and Mr. Vijay (PA to Prof. Kokate) for their assistance.

We also thank the management of Pharma Med Press for publishing the revised edition of the book.

C.K. Kokate

June, 2012 **S.B.Gokhale**

Contents

Chapter 1

General Introduction

The word **Forensic** is derived from Latin term Forencis, meaning a forum, a place for interaction or deliberations. **Jurisprudence** means study of fundamental laws and in case of pharmaceutical Jurisprudence, it is laws relating to pharmacy.

Forensic Pharmacy or Pharmaceutical Jurisprudence is that branch of pharmacy, which deals with various legislations pertaining to drugs and pharmaceuticals, and profession of pharmacy. This subject encompasses the knowledge of various Acts, Rules, Statutes, Schedules, Sections etc., which directly or indirectly influence the profession of pharmacy in the country and various operations pertaining to procurement, manufacture and distribution of different kinds of dosage forms. Three important components of pharmacy profession – education, industry and pharmacy practice – are regulated by the pieces of legislation enacted from time to time.

The knowledge of Forensic Pharmacy is essential to understand the legal aspects pertaining to practice of pharmacy. The qualified persons are required to profess and should also be engaged in manufacturing, sales and distribution of drugs. Pharmacy is a noble and dedicated profession with a commitment to the cause of health care system of the country. In order to ensure this professional role of pharmacist, there has to be an ethical framework within which a pharmacist is supposed to function. He/she should be familiar with the types of laws governing his/her profession and also the developments that have contributed to the current status of pharma education, pharmacy practice and pharmaceutical industry.

Since ancient times, the human race has been depending upon the plant-derived drugs for the treatment of different human diseases. Apart from our own civilization, the Chinese, Greek, Arabian and Tibetian civilization have contributed significantly to the knowledge of medicinal plants. In our country Ayurveda, the Ancient Science of Life, based on *'Tridosh'* theory of *Vaat* (wind), *Pitta* (bile) and *Kapha* (phlegm) is practiced from time immemorial. Our treatises or documents such as *'Vedas'* and *'Upanishads'* are full of information pertaining to medicinal plants. In ancient days, the medical care was in the able hands of *'Maharshis'* and *'Vaids'* who had a special status in the society. There was also the *Siddha* medicine mainly practiced in southern regions of the country.

With the advent of Moghul rulers specially Babur, there came in a new system of medicine practiced by Hakims called as Unani System of Medicine, which got patronage during the rule of Shahjahan and Aurangazeb. With the arrival of East India Company and other European companies and thereafter British rule in nineteenth century, the Indian population was first introduced to the Allopathic System of Medicine, more commonly known as "Vilayati Medicines". The modern system of medicine was introduced in India by the Dutch, the French, the Portuguese and East India trading companies and the Missionaries from European countries.

Until the end of the nineteenth century, the medicines of different systems were mostly derived from plants or other natural sources like animals and minerals. These drugs were in the form of extracts, tinctures, pills and pastes and most of them were freshly prepared. The Ayurvedic medical practitioners were mostly hereditary and they were following *Guru-Shishya parampara*, which was also true of Siddha and Unani practitioners (Hakims). The Homoeopaths were self-taught and relied mostly on literature from Germany. In the absence of legal requirements of registration as doctor, a large number of quacks surfaced in medical profession.

The hospital facilities were almost non-existent in rural areas. The railway administration and plantations provided good services to their employees. The Missionaries and charity hospitals for communities were serving limited cause of health care.

In British India, the European establishments like Kemp and Company; Bliss and Cotton; and Frank Ross and Company were the important pharmacies. The Indian companies in British India were: Popular Pharmacy at Bombay; Dadha and Company, Wilfred Perira Ltd., and Appah and Company at Madras; H.C. Sen and Company and The Young Friends and Company at Delhi; Beli Ram and Brothers, The Punjab Medical Hall and Narayan Das Bhagwan Das and Company at Lahore; and Butto Kristo Paul and Company and M. Bhattacharya and Company at Calcutta.

There were no legally controlled systematic manufacturing efforts in the country for the manufacture of different drug formulations to be used for a longer period. It was only when plant drugs were further processed/purified, and synthetic as well as semi-synthetic compounds of medicinal utility were manufactured and formulated in different dosage forms, the need to enact laws to govern various operations of manufacture, sale and distribution was acutely felt.

Current Status of Indian Pharmaceutical Industry

Indian pharmaceutical industry has undergone rapid metamorphosis in last one decade. The post-GATT era was the testing time for strategic planning and development of indigenous pharmaceutical sector. The mergers and acquisitions have become the order of the day. India has become hub for clinical trials. Millions of dollars are being invested in basic and applied R & D efforts. Quality of medicines has become the keyword for global survival.

Today India's position in world pharma market is 4th in terms of volume of medicines and 13th in value terms. Indian pharma companies are showing robust results and top ten Indian companies have recorded over 70% profits. There are around 24,000 manufacturing units, (large, medium and small scale) producing 90% requirement of medicines in the country. Currently, Indian companies produce about 22% of the world's generic drugs. The total drug production has already crossed mark of ` 95,000 crores of which anti-infective and cardio vascular drugs account for 14.7% and 11.1% respectively.

India is one of the top five Active Pharmaceutical Ingredients (API) produce in the world today. The emergence of Indian pharmaceutical industry on global scenario as a strong generic player is due to Indian

Patents Act, 1970, which allowed only process patents. Although, India shifted to product patent regime in 2005, the capabilities developed earlier have provided an edge over many other developing countries especially, when rising healthcare costs in many developed countries forced them to seek the cheaper drug option. In 2011, more than a third of the ANDA (abbreviated new drug applications) approvals in the U.S. were from Indian firms. The formulation exports from India, essentially generic drugs have grown at 21% compounded annual growth rate (CAGR) between 2005-06 and 2010-11. Around 150 billion US dollars worth of drugs are set to loose patent exclusively between 2010 and 2015, which would provide further boost to export from India.

The Indian biopharmaceutical industry is fast growing and presently, it is sized at about US $ 1.8 billion. It is yet to make inroads in U.S. and Europe. The R & D capabilities of Indian pharmaceutical industry currently, lie in reverse engineering drugs and in process chemistry. Traditionally, Indian companies have shied away from drug discovery, or in a few cases, out-licensed molecules to multinational companies at early stage of development. At present, a limited number of Indian companies are engaged in new drug research and around eighty molecules are in pipeline from Indian players. For sustaining the growth, Indian pharmaceutical companies will be required to look at newer avenues such as building relationships with global pharma for joint R & D activities, entering niche segments and widening distribution networks through marketing alliances. Other potential thrust areas include contract research, biopharmaceuticals and new drug research. Since the average cost of developing a new drug has more than doubled in last five years amounting to approximately US $ 1.5 billion, strategic partnership for new drug research is the need of the hour.

CHAPTER 2

History of Drug Legislation and Pharmacy Profession in India

During 1920-1930 there were a number of reports of harmful substitutes and adulterants being marketed in place of genuine drugs and toxic effects of such drugs were published in Indian press from time to time. According to reports of Indian Medical Gazette during this period, there was absolutely no control over the manufacture, sale and distribution of drugs in India. Several deaths were reported due to spurious drugs. In place of eye drops, croton oil was used. Chalk powder was frequently found to be used for adulteration of drug formulations. There were toxicity reports due to overdose of mercury compounds. In the absence of effective Acts and Rules related to drugs and pharmaceuticals in the country, there was a rat race for manufacturing of sub-standard, spurious and adulterated formulations.

As a result of alarming adverse reports, deaths due to spurious and adulterated drugs and in view of protests within and outside the country concerned with poor medical facilities offered by British rulers in India finally, on **9th March, 1927**, the British Government was forced to initiate action for drug legislations. The Council of States in British India headed by the Viceroy passed a resolution to counter or check malpractices in drug dispensation and medication. **On 11th August, 1930,** *Drugs Enquiry Committee* (D.E.C.) was constituted under the Chairmanship of Col. R. N. Chopra which was a historic development signaling beginning of new era of drug legislation in our country. Prior to the constitution of this committee, there was no significant piece of legislation regulating the import, manufacture, sale and distribution of medicine. No Act was in vogue prescribing qualification of a pharmacist and there was no systematic procedure adopted for registration as pharmacists.

Important Milestones in Drug Legislations and Pharmacy Profession (Education, Practice, and Industry)

PRE-INDEPENDENCE ERA

1664 The first hospital was opened at Fort St. George, Madras.

1811 Young Scotch named Mr. Bathgate came to India with East India Company and opened a Chemist's shop in Calcutta.

1820 Lord Cornwallis started opium factory at Ghazipur (U.P.).

1824 Hindustani versions (Devanagari and Persian scripts) of the London Pharmacopoeia were prescribed.

1824 The East India Company decided to impart knowledge of medical science—both European and Indian.

1835 First two medical colleges established at Calcutta and Madras.

1857 Few sections of Indian Penal Code were applicable for drugs.

1857, 1878 The Opium Act enacted.

1860 The beginning of pharmaceutical instructions in British India at Madras Medical College.

1868 The Pharmacopoeia of India published under the authority of Secretary of State for India.

1885 British Pharmacopoeia was made the sole authority for pharmacy profession.

1889 The Indian Merchandise Marks Act enacted.

1894 The Indian Tariff Act enacted.

1898 The Sea Customs Act enacted.

1899 The Compounders training course started in Madras.

1899 Acharya P.C. Roy along with Kartic Chandra Bose established Bengal Chemical and Pharmaceutical Works at Calcutta.

1905 Gajjar and Co., established at Bombay which also started drug manufacturing.

1906 In U.S.A. — Federal Food & Drugs Act introduced.

1919 The Poisons Act enacted.

1920 All India Compounders and Dispensers Association was established.

1920 In Canada — Food and Drugs Act introduced.

1924 The Cantonment Act enacted.

1925 In U.K. — The Therapeutic Substance Act introduced.

9-3-1927 Resolution of Council of States in India regarding health services.

1928 In U.K. Drug Adulteration Act enacted.

1928 The state medical faculty of Bengal introduced two years course for compounders.

11-8-1930 *Drugs Enquiry Committee* (D.E.C.) headed by Col. R. N. Chopra constituted.

1931 Report submitted by D.E.C. to Central Government.

1932 A two-year Degree Course in Pharmaceutical Chemistry for B.Sc. — Beginning of pharmacy education at Banaras Hindu University by Prof. Mahadev Lal Schroff (Father of Pharmacy Education in country).

1-11-1933 The Indian Medical Council Act enacted.

1935 United Provinces Pharmaceutical Association (UPPA) established at Banaras by Prof. Mahadev Lal Schroff.

1937 Import of Drugs Bill introduced in Parliament (British India) and later withdrawn due to criticism.

1937 Biological Standardization Laboratory (B.S.L.) established at Calcutta.

1939 United Provinces Pharmaceutical Association (U.P.P.A) was renamed as Indian Pharmaceutical Association (I.P.A). Publication of *Indian Journal of Pharmacy* started.

1940 Drugs Bill introduced in Parliament and Drugs Act later amended as **Drugs & Cosmetic Act (D.C.A)** was enacted.

1940 Indian Pharmacopoeal List.

1940 Biological Standardization Laboratory was named as Central Drugs Laboratory (CDL) under DCA.

1941 First Drugs Technical Advisory Board (DTAB) constituted.

1941 First All India Pharmaceutical Conference was held at B.H.U, Varanasi under the Presidentship of Prof. Mahadev Lal Schroff.

1943 *Health Survey and Development Committee* constituted under the chairmanship of Sir Justice Joseph Bhore.

1944 First I. P. Committee constituted.

1945 Pharmacy Bill introduced in Parliament.

1945 Justice Joseph Bhore submitted report.

1945 Rules for Drugs & Cosmetic Act framed.

POST-INDEPENDENCE ERA

1947 The Indian Nursing Council Act enacted.

1948 **The Pharmacy Act, 1948** enacted.

1948 The Dentists Act, 1948 enacted.

1949 First *'Pharmacy Council Of India'* (P.C.I.) constituted under the Pharmacy Act.

1949 Dr. K.C.K.E. Raja was nominated by the Central Government as the first President of Pharmacy Council of India.

1951 The Industries Act enacted.

11-7-1953 First Education Regulations (E.R) as approved by the Ministry of Health & F.W., Government of India were notified.

1954 The Drugs and Magic Remedies (Objectionable Advertisements) Act enacted.

1954 The Pharmaceutical Enquiry Committee recommended appointment of graduates in Pharmacy as Chief Pharmacists for all large hospitals.

1954 The first B. Pharmacy Course approved by Pharmacy Council of India at Birla College, Pilani.

1955 The first Diploma in Pharmacy Course approved by P.C.I. at Government Medical College, Amritsar.

1955 First Indian Pharmacopoeia published.

1955 The Medicinal and Toilet Preparations (Excise Duties) Act.

1956 Essential Commodities Act enacted.

1956 The University Grants Commission Act enacted.

1957 Dangerous Drugs (Import, Export & Transshipment) Rules framed.

1960 Prevention of Cruelty to Animals Act passed.

1960-70 Indian Drugs & Pharmaceuticals Ltd. (I.D.P.L.) established at five places in the country.

1962 The Central Manufactured Drugs Rules framed.

1962 Beginning of National Pharmacy week celebrations in third week of November every year.

1963 *Pharma Times* Publication of I.P.A as professional monthly publication.

1963 The Indian Hospitals Pharmacists Association (IHPA) was launched at Pilani, Rajasthan.

1964 *The Indian Journal of Hospitals Pharmacy* was started by Prof. B.D. Miglani for IHPA.

1966 Second Indian Pharmacopoeia published.

1968 Insecticides Act enacted.

1970 First DPCO (Drugs Price Control Order), later on in 1979 and 1987, 1995 published.

1970 Indian Patents Act enacted.

1971 Medical Termination of Pregnancy Act enacted.

1972 Education Regulations of P.C.I. 1972 (notified on 6-1-1973).

1973 Homoeopathy Central Council Act enacted.

1974 Committee of M.P.s with Jaisukhlal Hathi as chairman for drugs and pharmaceuticals constituted.

1975 Hathi Committee Report submitted. The Committee recommended that a Chief Pharmacist with at least a graduate in pharmacy degree should be appointed for maintaining quality of drugs supplied to patients in hospitals.

1975 All India Organization of Chemists and Druggists (AIOCD) established with Mr. V.L. Theagaraj as President .

1976 The Dentists (Code of Ethics) Regulations framed.

1977 Indian Pharmaceutical Congress along with Conference of Commonwealth Pharmaceutical Association was held under the Presidentship of Dr. J.N. Banerjee at Mumbai.

1978 Drug Policy was announced based on Hathi Committee report.

1979 *Indian Journal of Pharmacy* was named as *Indian Journal of Pharmaceutical Sciences* (Bi-monthly publication).

1981 Education Regulations of P.C.I. notified on 11.7.1992.

1985 Third Indian Pharmacopoeia published.

1985 The Narcotic Drugs & Psychotropic Substances Act enacted.

1986 Consumer Protection Act enacted.

1986 Revised Drug Policy was announced.

23-12-1987 The All India Council for Technical Education (AICTE) Act covering pharmacy education enacted.

1989 Golden Jubilee of Indian Pharmaceutical Association celebrated.

1991 Education Regulations of P.C.I. notified in 1993.

1994 Modifications in Drug Policy, 1994.

1996 Hon. Supreme Court directed Government to come out with National Policy on Blood Programme. Subsequently, National Blood Policy of NACO (National AIDS Control Organization) brought out.

1998 Golden Jubilee of Indian Pharmaceutical Congress alongwith Conference of Federation of Asian Pharmaceutical Association (FAPA) was held at Mumbai under the presidentship of Prof. C.K. Kokate.

1999 Golden Jubilee Year of Pharmacy Council of India celebrated throughout the country.

2001 The first pharmacist (Prof. C.K. Kokate) appointed as Vice-Chancellor of an Indian University (Kakatiya University A.P.).

2002 Pharmaceutical Policy announced by Ministry of Chemicals and Fertilizers, Department of Chemicals.

2005 In Post-WTO era, new patent regime (Product Patent) has started.

2008 Pharm. D and Pharm. D (Post-Baccalaureate) Programmes introduced by PCI.

2010 Diamond Jubilee Celebrations of Pharmacy Council of India at Delhi with President of India as Cheif Guest.

2011 Annual Conference of FIP (International Pharmaceutical Federation) was hosted by Indian Pharmaceutical Association at Hyderabad.

Drugs Enquiry Committee (D.E.C.), 1930

It was constituted in 1930 with Col. R.N. Chopra as Chairman, Shri C. Govindan Nair as Secretary and Dr. B. Mukherjee as Assistant Secretary. The other three members of the Committee were Fr.J.F. Caius, Mr. H. Cooper and Maulvi Abdul Matin Chowdhary. The Committee did splendid service to the cause of pharmacy profession in the country. The Committee visited many places in the country and interacted with the doctors and other professionals on the matters pertaining to health care system in country, quality of medicines available, etc. The questionnaires were sent by the Committee to medical personnel, customs officers, manufactures, medical associations and others related to medical field.

The **terms of references** of the Committee were as follows:

1. To probe into the quality of drugs that are being imported, manufactured and sold, especially those which are official in B.P.
2. To suggest remedial measures for checking import, manufacture, sale or distribution of the substandard or spurious drugs and their formulations.
3. To look into the formulations prepared indigenously from the vegetable drugs and suggest remedial measures for maintaining the quality of such formulations.
4. To look into all other aspects directly or indirectly connected with the profession of pharmacy.

D.E.C. reported that there was no systematic profession like pharmacy being practiced in the country. The drugs were dispensed and compounded or handled mostly by the untrained people. The remuneration paid to them was low. The so-called compounders with no knowledge of drugs were handling the drug formulations. They were also doing the work of dressers, helpers, laboratory technicians and all other miscellaneous jobs, including maintaining the accounts of doctors. These compounders were able to read and write English and that was the only qualification they had for handling the drugs. Only in the provinces of Bengal and Madras, there was a training course for compounders/ chemists & druggists. The report was a sad commentary on the poor state of the profession of pharmacy in the country. After an in-depth study and critical analysis, the Committee made the following recommendations to the Government:

1. There should be legislations to control drugs and other remedies whether belonging to the B.P or not.
2. There could be another law to ensure that drugs are handled by qualified persons and there could be a systematic course in pharmacy.
3. A drug regulating authority at center and in provinces/states be established.
4. There should be drug/quality control laboratories established in provinces and also at the centre. The efforts should be made to publish Indian pharmacopoeia.

The Committee submitted its report in 1931. The findings of the Committee spurred activity in teaching institutions, industry and the profession. Until 1937, British rulers did not act on this report. In 1937, *Import of Drugs Bill* was introduced with limited reference to the import and later withdrawn due to public criticism. Finally, in 1940 *The Drugs Bill* was introduced in the Parliament, based on the recommendations of D.E.C and after an in-depth deliberations *The Drugs Act,* 1940 was enacted which was later amended as the *Drugs and Cosmetics Act, 1940.*

The Health Survey and Development Committee, 1945, constituted under the Chairmanship of *Justice Bhore,* also re-emphasized the need for the qualified and trained pharmacists and registration of pharmacists, formation of Councils to govern the profession at centre and in provinces, strengthening of the provisions of Drugs Act, more drug control laboratories for strengthening of infrastructure for drug regulation, etc.

The recommendations of Drugs Enquiry Committee and Health Survey and Development Committee were responsible for laying the foundation for the *Pharmacy Act, 1948.* The Pharmacy Council of India was constituted in 1949 and the minimum qualification for registration as pharmacist was prescribed and process for registration described.

After Independence, it was felt to regulate the advertisements of drugs, which were in exaggerated form and misleading. A number of manufacturers were making exaggerated claims for their medicines and also exploiting the human weaknesses, especially in relation with advertisements pertaining to sexually transmitted diseases, menstrual disorders, loss of vigour, stamina, etc. The magic remedies were freely advertised for the cure of *Bhanamati*, epilepsy or fits, diabetes and a number of other diseases. The magic remedies in the form of Kavachas, Taits, Talisman, Sacred Bones, Sacred Bhasmas, Mantras, etc., were freely practiced and the poor and illiterate people were exploited. It is to control such objectionable trends, "*The Drugs & Magic Remedies* (Objectionable Advertisements) *Act*" was passed by the Parliament in 1994.

Alcohol is an important solvent in pharmaceutical industry. Alcohol is required for manufacturing of drugs and also for drug formulations as vehicle, preservative and therapeutic agent. At the time of Independence, manufacturing, sale and distribution of alcohol was controlled by provinces/states. The rates charged by the provinces/states in the form of excise duties were different for various regions. Alcohol was also being misused and it was a drug of abuse. It is to regulate the production, sale and distribution of alcohol and to bring uniformity in the excise duties to be paid, the *Medicinal and Toilet preparations* (Excise Duties) *Act* was passed by the Parliament in 1955.

The Drugs Price Control Order, 1970 was aimed at fixing the prices for the drugs and their formulations categorized into essential and non-essential groups so that uniform retail rates for certain categories of drugs can be maintained and the drugs prices of life-saving drugs can be controlled.

The Indian Patents Act 1970, which was process patent-based, gave an impetus to Indian pharmaceutical industry which resulted increased indigenous production of drugs and pharmaceuticals. In post-WTO era, with the new product patent regime coming into force, the Indian pharmaceutical industry is undergoing rapid metamorphosis to meet the challenges of globalization of pharmaceutical trade. Many Indian pharmaceutical companies have acquired the status of multinational companies. A new era of pharmaceutical profession in India has just begun to meet the challenges of Post-GATT and Post-GATS developments.

CHAPTER 3

Pharmaceutical Ethics

Pharmacy is a noble and sacred profession committed to the cause of health care of human beings. The pharmacist is a vital link between the doctor and the patient. He is charged with the responsibility of providing professional services of high order to the community at large by ensuring production of quality medicine and their sale and distribution to the consumers.

The "Laws" refer to the rules for conduct of human beings, binding on all citizens in a state or the nation. While "Ethics" cover the set of rules by which a profession regulates action and sets standards for its members.

Pharmaceutical ethics may be defined as "rules of conduct or standards by which pharmacy community regulates its actions and sets standards for its members". The pharmaceutical ethics encompasses the code of moral principles or the science of morals which is concerned with human character. The pharmacist of today is a drug-maker, drug-dispenser, drug-custodian, patient-counsellor, drug-researcher and drug-educator, and above all an honest and patriotic citizen. The techno-professional background of the pharmacist gives him/her the confidence of providing services with ethical approach to the satisfaction of patients. The sacred values are required to be cherished and professed by the pharmacist.

Government restricts the practice of pharmacy to those who qualify under regulatory requirement and expects the pharmacist to fulfill his/her professional obligations honourably and with due regard for the well-being of the society.

Pharmacist's Oath

(As approved by the Pharmacy Council of India)

I swear by the Code of Ethics of Pharmacy Council of India in relation to the community and shall act as an integral part of health care team.

I shall uphold the laws and standards governing my profession.

I shall strive to perfect and enlarge my knowledge to contribute to the advancement of pharmacy and public health.

I shall follow the system which I consider best for pharmaceutical care and counseling of patients.

I shall endeavour to discover and manufacture drugs of quality to alleviate sufferings of humanity.

I shall hold in confidence the knowledge gained about the patients in connection with my professional practice and never divulge unless compelled to do so by the law.

I shall associate with organizations having their objectives for betterment of the profession of pharmacy and make contribution to carry out the work of those organizations.

While I continue to keep this oath unviolated, may it be granted to me to enjoy life and the practice of pharmacy respected by all, at all times!

Should I trepass and violate this oath, may the reverse be my lot!

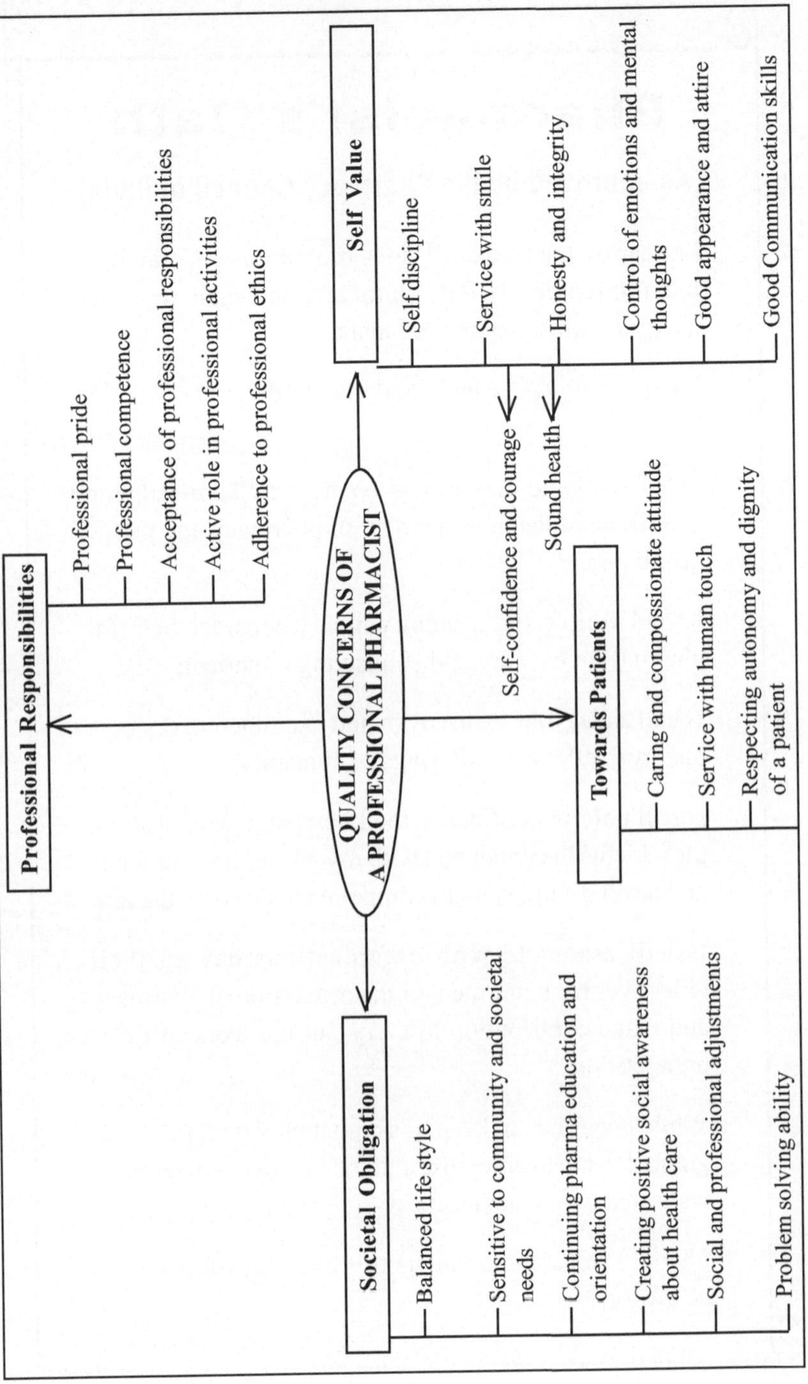
QUALITY CONCERNS OF A PROFESSIONAL PHARMACIST
Professional Responsibilities
Professional pride
Professional competence
Acceptance of professional responsibilities
Active role in professional activities
Adherence to professional ethics
Self Value
Self discipline
Service with smile
Honesty and integrity
Control of emotions and mental thoughts
Good appearance and attire
Good Communication skills
Self-confidence and courage
Sound health
Towards Patients
Caring and compossionate attitude
Service with human touch
Respecting autonomy and dignity of a patient
Societal Obligation
Balanced life style
Sensitive to community and societal needs
Continuing pharma education and orientation
Creating positive social awareness about health care
Social and professional adjustments
Problem solving ability

Principles of Pharmaceutical Ethics

- The pharmacist should be a good and law-abiding citizen conversant with Acts and Rules governing his/her profession.
- He/she should be a person with an attitude of service and sacrifice and concern for welfare of both patients and public.
- The pharmacist must be capable of upholding honour and dignity of the profession and not getting involved in any action which may bring disrepute to the profession.
- The pharmacist should never disclose any information about patient or his family which he/she has acquired as part of professional activities, unless required to do so by law.
- The pharmacist should not get involved in any unethical work under duress or pressure.
- Unethical and cut-throat publicity of professional services offered by the pharmacist is against basic values of the profession. The pharmacist should not get involved in such unethical practices.
- The pharmacist should offer services directly to the public in his/her own premises which reflect the professional character of pharmacy. He/she should provide efficient and reasonably comprehensive service.
- The pharmacist should receive the prescription of a doctor without any comment or discussion over it. He/She should not add, omit or substitute any content of the prescription without the consent of the prescriber. He/She should answer any questions on the prescriber to the customers with caution and core. In case of any error in the prescription, it should be referred back to the prescriber for correction.
- The pharmacist should keep himself/herself abreast of the progress of pharmaceutical knowledge in order to maintain high standard of professional competence. He/She should exchange such information with fellow-pharmacist.
- The pharmacist should resist the temptation of recommending any particular medical practitioner, unless specially asked for.

- The pharmacist as an integral component of health care team should always endeavour to cooperate with doctors, nurses and other members of health care system.
- The pharmacist should possess good communication skills to be able to work closely with other health care providers.
- The pharmacist should help the client with personal consultation and about proper use of medicines and other health care products. The information provided must be clear, understandable, correct, up-to-date and explicit.
- The pharmacist should be alert to occurrence of adverse effects of medicine and same should be recorded in the individual patient medication record.
- The pharmacist should be an active member of all professional associations or organizations which are established with an objective of sincerely serving the cause of the profession.

The chariot of health care system is driven by four giant wheels represented by doctors, pharmacists, nurses and other paramedical personnel. So long as the four wheels of the chariot are co-axial, the chariot of health care shall march steadily on the path of prosperity and good health. The moment there is imbalance in mobility of the chariot due to defect in one of the wheels, the chariot is bound to collapse mid-way, and the dream of attaining excellence in health remains unfulfilled.

CHAPTER 4

The Pharmacy Act, 1948

The recommendations of Drugs Enquiry Committee, as also Health Survey and Development Committee, laid the foundation for the enactment of the Pharmacy Act, 1948. The necessity was felt to monitor the profession of pharmacy in the country in order to ensure that the medicines are handled, dispensed/compounded or sold only by the qualified persons, namely, the registered pharmacists.

In the absence of any uniform regular course for the training of pharmacist in the country in the pre-independent era, the necessity was also felt to introduce a professional course of instruction as minimum qualification of registration as pharmacist, in order to create requisite qualified manpower to manage the affairs of profession of pharmacy in the country.

The drugs are life-saving and also potent which require proper handling at the counters in retail outlets and hospitals. Their mishandling by non-qualified persons shall play havoc with national health care system. It is to overcome this obstacle the Pharmacy Bill was introduced in 1945.

On 4th March, 1948, the statutory control on the pharmacy education in the country was established with the enactment of the Pharmacy Act, 1948 with the following preamble.

"An Act to regulate the profession of pharmacy". Whereas it is expedient to make better provision for the regulation of the profession and practice of pharmacy and for that purpose to constitute Pharmacy Councils;

It extends to whole of India except the state of Jammu and Kashmir.

The Pharmacy Act enacted by Ministry of Health and Family Welfare, Government of India, is covered under five Chapters which encompass 46 sections. The major amendment to the Act was made in 1976 to Section 42 of the Act wherein after cut-off date of 1st November, 1983, the drug stores in the country would be run under the supervision of registered pharmacists.

The chapters covered under the Pharmacy Act are as follows:

Chapter	I	Introductory
Chapter	II	Pharmacy Council of India (PCI)
Chapter	III	State Pharmacy Council (SPC)
Chapter	IV	Registration of Pharmacists
Chapter	V	Miscellaneous

Chapters I and II came into force immediately on enactment of the Act. Chapters III, IV and V were to be implemented within the timeframe given by the Central Government to the State Government by publication in the Official Gazette or the respective Union Territory. First Pharmacy Council of India was constituted in 1949. First Education Regulations (E.R) were to be framed by 1952 and effectively implemented within 3 years of their framing.

Pharmacy Council of India

On 9th March, 1949, the Pharmacy Council of India (PCI) was constituted to fulfill the objectives of the Pharmacy Act, 1948 by way of:

1. Prescribing the minimum standard of education required for qualifying as a pharmacist i.e., framing of Education Regulations prescribing the conditions to be fulfilled by the institutions seeking approval of the PCI for imparting education in pharmacy.
2. Ensuring uniform implementation of the educational standards throughout the country.
3. Approving the courses of study and examination for pharmacists i.e., approval of the academic training institutions providing pharmacy courses.

4. Withdrawing approval, if the course of study does not continue to be in conformity with the educational standards prescribed by the PCI.
5. Approving qualifications granted outside the territories to which the Pharmacy Act, 1948 extends i.e., the approval of foreign qualification.
6. Maintaining Central Register of Pharmacists.

The composition of Central Council of PCI is given in Table 4.1.

The Central Council of PCI elects the President, Vice-President and five members of the Executive Committee (EC) from amongst its members by election conducted in accordance with the norms laid down in the Pharmacy Act. Any dispute arising out of the election of Executive Committee of PCI is referred to the Ministry of Health and Family Welfare, Government of India, whose decision shall be final.

The President or in his absence Vice-President chairs the meetings of the Central Council and the Executive Committee. The Central Council meets at least once in a year, whereas the Executive Committee meets more frequently depending on the quantum of work usually, four times in a year. The term of the President, Vice-President and the Council Members is for five years unless otherwise, re-elected or renominated, as the case may be. Ex-officio members of PCI hold their membership by virtue of their designation and cease to be the members on retirement or replacement. The Director General of Health Services, Government of India, and the Drugs Controller General of India may send their nominees in writing for attending a specific meeting in case they are unable to attend the meeting.

Any member of the Central Council or Executive Committee, who remains absent continuously for three meetings without intimation or without any valid reason, is eligible to be disqualified as a member of the Council.

The Council appoints the Secretary or Registrar of PCI, who can also act as Treasurer. The Council appoints the other staff including Deputy Secretary or Assistant Secretaries to run the office.

Table 4.1 Composition of Central Council of PCI.

Ex-Ofmficio **Members**	**Elected Members**	**Nominated Members**
(i) Director-General of Medical and Health Services, Government of India or his/her nominee.	(i) Six members of whom at least one teacher of each of the subjects of Pharmacy, Pharmaceutical Chemistry, Pharmacology and Pharmacognosy elected by the University Grants Commission from amongst teacher of Indian Universities or affiliated colleges imparting diploma or degree in pharmacy.	(i) Six members nominated by the Government of India, Ministry of Health and Family Welfare, of which at least 4 should possess degree or diploma in pharmacy qualification and should be practicing pharmacy or pharmaceutical chemistry.
(ii) Drugs Controller General of India or his/her nominee.	(ii) One member elected from amongst its members of Medical Council of India.	(ii) One representative each of the University Grants Commission (UGC) and All India Council for Technical Education (AICTE).
(iii) Director, Central Drugs Laboratory (CDL), Kolkata.	(iii) One member each elected by the State Pharmacy Council of each State who should be a registered pharmacist.	(iii) One member representing each State Government or Union Territory who shall be a registered pharmacist nominated by the respective State Government or Union Territory.

The Central Council elects Chairmen of Education Regulations, Professional Pharmacy and Law Committees. The President and Vice-President of PCI are Ex-officio members of these sub-committees.

The Council is a corporate body, a statutory organization which can sue or be sued. The Council can hold movable and immovable property and maintain accounts and registers. The minutes of the meetings of Central Council are required to be sent to Ministry of Health and Family Welfare, Government of India, and all the decisions of the approval or otherwise of the institutions or colleges are published in the Official Gazette.

Minimum Qualification for Registration as a Pharmacist in India

The minimum qualification for registration as a pharmacist in India is a pass in "Diploma course in pharmacy" from an institution recognized by the PCI. Persons holding "Degree in Pharmacy" from an institution recognized by the PCI are also eligible for registration as a pharmacist. Section 10 of the Pharmacy Act empowers the PCI to frame regulations called Education Regulations (E R) prescribing the minimum standard of education required for qualification as a pharmacist. These regulations have been revised from time to time to keep pace with the latest developments in the field of pharmaceutical sciences.

The PCI has decided to upgrade the minimum qualification for registration as a pharmacist from Diploma to Degree in Pharmacy and this decision of the PCI is under consideration of the Ministry of Health and F.W., New Delhi.

Education Regulations Framed by the Pharmacy Council of India under Section 10 of the Pharmacy Act, 1948

Education Regulation (ER) basically prescribes-

1. The standard of education for qualification as a pharmacist i.e. minimum qualification for admission to the course, duration of the course, the syllabus, mode of examination, minimum marks for passing the examination, nature and duration of practical training, etc.

2. Minimum conditions which an institution has to provide for seeking approval of the PCI for conducting a course of study for pharmacists.
3. conditions to be fulfilled by the Examining Authority for approval for conduct of examinations.
4. Conditions to be fulfilled by the institutions to be recognized for giving the practical training.
5. Practical training contract forms for the pharmacists.

The PCI is charged with the responsibility of regulating profession of pharmacy in the country by way of prescribing Education Regulations which are required to be followed by any institution imparting D. Pharm. Education. The Education Regulations are also required to be covered by the institution imparting B. Pharm. Programme.

PCI appoints a set of inspectors called Pharmacy Inspectors who are required to inspect the institutes periodically preferably at the time of examination for the physical facilities provided in terms of laboratories, classrooms, teaching and non-teaching staff structure, teaching instructions, attendance of students and staff members, time table, records of sessional marks, conduct of annual exams, assessment of the papers in sessional and annual examination, etc. On the instructions of the President of PCI, these inspectors usually two for one institute conduct the inspection on the basis of SIF (Standard Inspection Form) submitted by the institution. The Executive Committee goes through the inspection report and takes the decision on the recognition of the institution for a fixed duration subject to the approval of the Central Council. If there are serious deficiencies pointed out in the report, the Executive Committee asks for the compliance from the institution within three months. If the institution fails to send a compliance report the Council, PCI can serve a withdrawal notice, to the institution for the withdrawal of recognition. If convincing reply is not sent by the institution through the State Government or examining authority within stipulated period after serving of the notice, the Council can withdraw recognition and such a withdrawal is published in Official Gazette. No pharmacy institution imparting Diploma can exist in the country without the approval of PCI. Any person who holds D.pharm/B.pharm is eligible to register as a registered pharmacist with the State Pharmacy Council concerned after making necessary payments of fees and submission of evidence of qualification from an institution approved by PCI.

The institution which is de-recognised by PCI can apply afresh for running the course after providing all necessary facilities as per ER. The Council can arrange for inspection once again.

Education Regulations 1991 cover different requirements for Diploma in Pharmacy Part I, II, and III.

Diploma in Pharmacy Part I and Part II are the actual instructions in I and II year, respectively, whereas Part III covers practical training.

First year of D. Pharm

The subjects covered are:

(i) Pharmaceutics – I	(Theory and Practical)
(ii) Pharmaceutical chemistry – I	(Theory and Practical)
(iii) Pharmacognosy	(Theory and Practical)
(iv) Human Anatomy and physiology	(Theory and Practical)
(v) Biochemistry and clinical pathology	(Theory and Practical)
(vi) Health Education and community pharmacy	(Theory)

Second Year D. Pharm

The subjects covered are:

(i) Pharmaceutics –II	(Theory and Practical)
(ii) Pharmaceutical chemistry – II	(Theory and Practical)
(iii) Pharmacology and toxicology	(Theory and Practical)
(iv) Hospital and Clinical pharmacy	(Theory and Practical)
(v) Pharmaceutical jurisprudence (Forensic pharmacy)	(Theory)
(vi) Drug store and Business management	(Theory)

Requirements as per ER 1991

1. Physical Facilities

For an intake of 60 in first year D. Pharm at least, four laboratories, one each for pharmaceutics, pharmaceutical chemistry, biochemistry and clinical pathology, and pharmacognosy/APH.

The classrooms (at least two), aspectic room, balance room, animal house, drug-museum, machine room, Principal's office, girls and boys common rooms, etc., should be provided. The laboratory should be minimum 500 sq. ft. in area. Equipments, instruments and glassware as given in E R should be provided.

2. Staff

(i) Principal/Director/Head of Department should be of the rank of professor or reader with M. Pharm. and minimum of five years teaching experience.

(ii) Senior lecturer or lecturer: M. Pharm in the subject concerned or B. Pharm with three years professional experience.

For the subjects like clinical pathology, toxicology, pharmacology, clinical pharmacy even MBBS qualification is permitted.

3. Instructions

(i) A minimum of 180 working days including time spent for sessional exams.

(ii) In a week, a minimum of 36 hours of teaching, preferably six hours per day.

(iii) Staff/Student ratio in theory classes 1: 60

(iv) Staff/Student ratio in practicals 1 : 20. A practical batch should not have more than 30 students wherein two staff members are required; for a batch up to 20 student, one staff member is needed.

(v) There should be not more than two classes in the same subject by particular teacher on the same day.

(vi) A minimum of 75 percent attendance is compulsory, separately, in theory and practicals.

4. Examinations

There should be at least two sessional exams. Average of best two is taken into consideration. Each paper carries 100 marks of which 20 are for sessional and 80 for annual examination. In case of the practical sessional examination of 20, ten marks are for the day-to-day work and record and other ten marks for actual performance in examinations.

Under E.R 1991 the minimum passing marks are 40 percent in each paper inclusive of sessional marks. A candidate securing 60 percent or more and less than 75 percent is declared to have passed in first class. The distinction is awarded to a candidate scoring 75 percent or more.

After having passed D. Pharm Part II examination, the student has to undergo practical training in PC-approved hospital/dispensary/ clinic/drug store/ drug unit/charity hospital or any other institute. The practical training is for a minimum of 500 hours spread over not less than three months. Of these, at least 250 hours should be devoted for dispensing/compounding of drugs or handling of drug formulations. During the course of training, the student shall work under the guidance of **Apprentice Master** who is a registered pharmacist. The student shall be trained in drug laws, dispensing of drugs, maintaining records, identification of drugs from different sources and all other related matters. The training form is in five parts.

Part I : It is to be filled by the Head of the Institution (College).

Part II : It is to be filled by the Trainee (student) as an understanding.

Part III : It is to be filled by Apprentice Master for accepting the student for training.

Part IV : It is to be filled by the student after training.

Part V : It is to be filled by the Institution certifying that student has undergone practical training as per ER 1991.

This certificate along with D.Pharm. pass certificate and required fee is to be submitted to the State Pharmacy Council concerned for the purpose of registration as pharmacist.

Pharm.D and Pharm.D (Post Baccalaureate)

The profession of pharmacy is undergoing rapid metamorphosis and witnessing tremendous technological advancements. It is now moving forward in the pharmaceutical care sector with focus on patient. With change in disease pattern and dramatic increase in number of medicines, the challenges ahead are related access to quality medicines and their rational use. The scope of pharmacy practice has shifted from dispensing of medicines to patient counseling, drug information, monitoring drug therapy etc.

In 2008, PCI has introduced Pharm.D course after extensive curriculum revision and restructuring. The syllabus lays emphasis on pharmacy practice. The role of pharmacist in healthcare, Professionalism, dispensing, health and the individual and health care in our society. Pharmacoepidemiology, the psychosocial aspects of medicine treatment, communication skills, patient counselling, and legal aspects of practice are included in the course content. During the fifth and sixth year, students undertake professional experience placement in hospitals for onsite bed situational training on live disease patterns from active to chronic. The course has an orientation for clinical aspects of medicines in terms of patient safety.

Pharm.D course is a professional pharmacy doctoral programme of six years duration after 10 + 2 (science stream) which includes, five years of academic study and one year of internship or residency.

The eligibility criteria for admission in to Pharm.D course

1. A pass in 10 + 2 examination with physics and chemistry as compulsory subjects along with, either mathematics or biology or
2. A pass in D.Pharm course from an institution approved by the Pharmacy Council of India under section 12 of the Pharmacy Act.

The eligibility criterion for admission into Phnrm.D (Post Baccalaureate) course

1. A pass in B.Pharm from an institution approved by the Pharmacy Council of India under section 12 of the Pharmacy Act.

Number of admissions for the Phann.D program shall be as prescribed by the Pharmacy Council of India from time to time and presently, it is restricted to thirty students intake for Phann.D and ten students for Pharm.D (PB) per academic year. Pharm.D and Pharm.D (PB) are registrable qualifications under the Pharmacy Act, 1948.

Under Pharm.D Regulations, there is a mandatory requirement of signing a MOU with 300 bedded corporate hospital/teaching medical institution. Since part of training also includes participation of pharmacists in monitoring prescription in out-patient departments, running drug information centres, contribution in drugs and therapeutics committees, etc., the PCI is utilizing this opportunity to sensitize the other health professionals about pharmacist's role in the health care.

It is opined that the introduction of Pharm.D nd Practice Regulations with emphasis on clinical practice, shall be in line with the proclivity in health care standards in this challenging scenario. This optimizes the contribution of pharmacists to pharmaceutical and patient health care and the pharmacist will be able to provide trustworthy drug information to both patient and physician. This will open new doors for patient counseling besides, assisting physician about medicines.

State Pharmacy Council (SPC)

State Pharmacy Council is charged with the responsibility of maintaining an up-to-date register of the registered pharmacists within the State. With an amendment to Section 42 of the Pharmacy Act, no person other than registered pharmacists can compound, dispense or do the retail business pertaining to medicines.

State Pharmacy Council of one State can also cater to the needs of other States through an official agreement. There can also be Joint State Pharmacy Council covering two or more States under an official agreement.

The composition of State Pharmacy Council constituted by the State Government or the Joint State Pharmacy Council constituted by participating States is as follows:

State Pharmacy Council	Joint State Pharmacy Council
I. Elected Members	
(i) Six members elected by registered pharmacists of the particular State.	(i) 3-5 members elected from Registered pharmacists of each participating State.
(ii) One member who is a registered medical practioner elected by the Medical Council of State.	(ii) One registered Medical Practioner elected by the Medical Council of each participating State.
II. Nominated Members	
(iii) Five members nominated by each State of which at least three should be Degree or diploma holders in pharmacy or pharmaceutical chemistry.	(iii) 2-4 members nominated from each participating State of which more than 50% shall be the persons with degree or diploma qualification in pharmacy or pharmaceutical chemistry.
III. Ex-Officio Members	
(iv) (a) Chief Administrative Office or Incharge, Medical and Health Services of State or his/her nominee. (b) Officer Incharge for Drugs and Cosmetic Act, 1940 of each State, or his/her nominee. (c) One Government Analyst under DCA, 1940 nominated by the State Government.	(iv) (a) Chief Administrative Officer or Incharge, Medical and Health Services of each participating State or his/her nominee. (b) Officer Incharge for Drugs and Cosmetics Act, 1940 of participating State or his/her nominee. (c) One Government Analyst under DCA, 1940 nominated by each participating State.

The term of State Pharmacy Council is for five year. The Council elects President and Vice-President of State Pharmacy Council who hold office for five years, provided he/she continues as a member of Council. The Executive Committee of State Pharmacy Council (SPC) comprises five members elected from amongst its members. The SPC appoints a Registrar who shall be the Secretary of the Council and he/she may also be the Treasurer. The appointment of other officers, fixing of the salaries and payment of allowances to the members are also done by SPC. The SPC is statutory body with perpetual succession, can hold property with a common seal and sue or be sued. State Pharmacy Council is custodian of the register of registered pharmacists in the State. The Council sends five copies of register to Pharmacy Council of India. State Pharmacy Council can fix the registration fee on annual basis or a long-term basis. It has to send 1/4th of its collection of registration fee every year before 1st of May to Pharmacy Council of India. State Pharmacy Council, from time to time, keeps informed the State Government concerned about its activities. The SPC also organizes continuing education programmes for registered pharmacists with a view to updating their professional skill and knowledge.

A State Pharmacy Council, with permission of the Government, may appoint Inspectors with prescribed qualification under section 26-A to inspect any premises where drugs are compounded or dispensed, enquire the qualification of person engaged in professional activity, investigate complaint in writing and institute prosecution in consultation with SPC.

Registration of Pharmacist

1. **First Register of Pharmacist:** Immediately after Independence until SPC was constituted.
2. **Subsequent Register:** During the period of implementation of Pharmacy Act and framing of Educational Registrations.
3. **Regular Register:** After Education Regulations came into force.

1. First Register of Pharmacist

Immediately after Independence, there was shortage of pharmacists to run the hospitals and drug stores. Due to chaotic condition prevailing then, the State Governments were asked to appoint *Registration Tribunal* of three members of which one shall act as Secretary-cum-Treasurer. In many States, this Registration Tribunal consisted of professional and judiciary persons. Registration Tribunal fixed last date for submitting application for registration as pharmacists, provided he/she fulfilled following qualifications.

1. Minimum age of eighteen years on date of application and
   (i) If he/she holds degree/diploma in pharmacy or pharmaceutical chemistry or an approved Chemist and Druggist diploma issued by any State Government or a prescribed qualification granted by an authority outside India.

Or

(ii) A graduate in any discipline with a minimum of three years experience in compounding and dispensing in a hospital or dispensing at any other place where drugs are regularly dispensed on prescriptions of medical practitioners.

Or

(iii) A person who has passed examination recognized as adequate by the State Government for compounders or dispensers.

Or

(iv) A person with minimum of five years experience in compounding or dispensing in a hospital or dispensary on prescriptions of medical practitioners.

After receipt of applications, the Registration Tribunal scrutinizes the same and directs Secretary to enter the names of eligible candidates in Register as pharmacists. Any person aggrieved or not satisfied with the decision of Registration Tribunal was given 60 days time to appeal to a Special Appelate Authority constituted for this purpose by the State Government and the decision of Appellate Authority was final and binding. After the formation State Pharmacy Council, the register was handed over to it.

2. Subsequent Register

For the interim period until Education Regulations were enforced in the State, the names of eligible candidates were included in the register of pharmacists on fulfilling the following criteria.

1. Minimum age of eighteen years
2. A person satisfying conditions prescribed with the prior approval of the Central Council or conditions entitling a person to have his name entered on the first register, provided he is a matriculate.

Or

A person with matriculate qualification who is a registered pharmacist in another State.

3. Regular Register

After implementation of Education Regulations framed by the Pharmacy Council of India, only a diploma in pharmacy holder of an institution approved by PCI who has undergone practical training of 500 hours after completion of D. Pharm. course is only eligible for registration as pharmacist.

There were two major amendments to the Pharmacy Act, 1948 for the purpose of registration of pharmacists in 1959 and 1976.

According to the amendment to Act in 1959, the First Register was once again opened for consideration of following categories of people for the purpose of registration as pharmacists.

(i) The persons who were displaced as a result of Partition of country from erstwhile territory of India which is part of Pakistan and who were carrying on the business or profession of pharmacy as their principal means of livelihood on or before **4th March, 1948,** and satisfy the conditions for inclusion in the First Register of Pharmacists.

(ii) The citizens of India who have been carrying on the business or profession of pharmacy in any country outside India, provided they are qualified for registration in First Register.

(iii) The persons who resided in an area which has become a territory of India and they satisfy the conditions for registration in First Register.

(iv) The persons belonging to an area which has become part of other State as a result of State Reorganization on **1st November, 1956.** Even though, the First Registration in such a State was closed, the persons in this category were enrolled as pharmacists in First Register provided they have fulfilled the criteria laid down for enrollment in the First Register.

(v) A person who has not responded to the call of registration in First Register even though he/she was qualified to be registered, due to ignorance or any other reason, was once again given one-time opportunity for registration.

The **amendment** to the Pharmacy Act in **1976** had considered following categories of people for registration as pharmacists:

(i) The displaced persons as a result of Bangladesh war or from Sri Lanka, Burma and Uganda or any other country between **14th April, 1957** and **25th March, 1971,** provided they were carrying on business or profession of pharmacy as their principal means of livelihood for a period of not less than five years. This clause was also applicable to the persons affected by Goa Liberation Movement who were practicing pharmacy as per requirement in First Register.

(ii) The persons identified in the provision for the Drugs and Cosmetic Act, 1940 as qualified persons until **31st December, 1969** who were professing pharmacy were also considered for registration as pharmacists.

These amendments were necessitated on humanitarian grounds and to attend to genuine difficulties in preparing the First Register. However, as on today, the registration of pharmacists is strictly for diploma or degree holders in pharmacy from an institution approved by PCI.

The State Pharmacy Council maintains up-to-date register of pharmacists after collection of requisite fees including following information.

(i) Full name of pharmacist and his/her residential address
(ii) The date of his/her first admission to the Register
(iii) Qualification, year of passing of qualifying examination
(iv) Professional address
(v) Date of birth and any other particulars as prescribed.

The Registrar appointed by SPC makes necessary entries in the Register and issues the receipt for registration which is a documentary evidence for registration. A certificate of registration is also issued by State Pharmacy Council on payment of fees. Anyone who is aggrieved by the decision of Registrar for non-inclusion of his/her name in Register may appeal to SPC within three months and the decision of SPC regarding registration shall be final.

The Registrar, on payment of fees, may issue duplicate certificate only on confirmation that the original is lost or destroyed.

Removal of Name from Register

Subject to provisions of Section 36 of the Act, the Executive Committee may order for removal of name of a person from Register, after giving reasonable opportunity to person concerned, on following grounds:

(i) A person whose name has been entered in the Register by error or an account of misrepresentation or suppression of fact,

(ii) A person who gives false information about himself/herself,

(iii) A person who submits false certificate or false document in support of his/her registration,

(iv) A person who is convicted of an offence in connection with his/her profession,

(v) A person who was indirectly involved in commitment of professional offence and if it is proved that the offence was instigated or connived at by the registered pharmacist.

A person aggrieved by order of State Pharmacy Council for removal of name from Register may appeal to the State Government within one month and the order of the State Government upon such appeal shall be final.

A person whose name has been removed from the Register shall surrender forthwith his/her certificate of registration to the Registrar of SPC and the name so removed shall be published in the Official Gazette.

Offences and Penalties

Dispensing by Unregistered Person: Under Section 42 of the Act, no person other than a registered pharmacist shall compound, prepare, mix, or dispense any medicine on the prescription of a medical practitioner. Whoever contravenes this provision is punishable with imprisonment of six months or fine of ₹ 1000.00 or both. This shall not apply to the dispensing by a practitioner for his own patients.

The penalty for falsely claiming to be registered pharmacist is ₹ 500.00 for first conviction and imprisonment extending to six months or fine of ₹ 1000.00 or both for subsequent conviction.

Appointment of Commission of Enquiry

If the Central Government has any substantial reason to believe that Pharmacy Council of India is violating the norms laid down in the Pharmacy Act, 1948 or misusing the powers or indulging in the matters which are not in accordance with Statutes pertaining to the Act, the Central Government is empowered to constitute Commission of Enquiry to probe into the affairs of Pharmacy Council of India. The Commission of Enquiry of three persons, two of whom shall be appointed by the Central Government, one being the judge of a High Court and the other nominated by the Council shall enquire into the matters referred to it and submits its report or recommendations to the Central Government. The Government may accept the report or remit the same to the Commission for modification or reconsideration. After the report is finally accepted, the Central Government orders PCI to comply within specified time, failing which, the Central Government may take appropriate action.

At State Government level, the State Government constitutes Commission of Enquiry for State Pharmacy Council not functioning in accordance with provisions of the Act and passes such order or take necessary action as deemed fit.

CHAPTER 5

The All India Council for Technical Education Act, 1987 (AICTE Act, 1987)

Salient Features

The Act envisages proper planning and coordinated development of technical education system in the country and promotion of qualitative improvements of technical education in relation to planned quantitative growth. The *"technical education"* covers programmes of education, research and training in engineering, technology, pharmacy, architecture, town planning, management and applied arts and crafts. The Act was enacted on 15th December, 1987. It has six Chapters and 25 Sections.

Council

The Council of AICTE is a body corporate with perpetual succession and common seal and shall by the said name sue and be sued. The head office of the Council is at Delhi. The Council is represented by 51 members of which three are AICTE officials, fifteen Ex-officio, 31 nominated members and two Members of Parliament (elected)

1. Chairman
2. Vice-Chairman
3. Member-Secretary

} Appointed by the Central Government

4. Ex-officio members (fifteen members)
   (i) Secretary of Education, Govt. of India
   (ii) Educational Adviser (General), Govt. of India

(iii) Chairmen of Four Regional Committees of AICTE
(iv) Chairmen of Five All India Boards of AICTE
(v) Chairman, University Grants Commission
(vi) Director, Institute of Applied Manpower Research
(vii) Director-General, Indian Council of Agricultural Research
(viii) Director-General, Council of Scientific and Industrial Research

5. Members appointed by the Central Government (Total 31)

    - One representative of Ministry of Finance, Government of India.
    - One representative of Ministry of Science and Technology, Government of India.
    - Four members appointed by rotation to represent Ministries and departments of Central Government other than Finance and Science and Technology.
    - Four members representing organizations in industry and commerce.
    - Eight members appointed by rotation in alphabetical order to represent States and Union Territories.
    - Seven members to represent Central Advisory Board of Education; Association of Indian Universities; Indian Society for Technical Education; Council of Indian Institutes of Technology; Pharmacy Council of India; Council of Architecture; and National Productivity Council.
    - Four members to represent professional bodies in technical and management education and maximum two members appointed to represent other areas.

6. Two Members of Parliament of whom one shall be elected by the House of the People and one by the Council of States.

The term of the office, other than ex-officio members, is for five years. There is a provision for filling up the vacancies and co-option of persons for particular purposes. The Council usually meets twice a year. The Council takes decisions pertaining to development of technical education; allocation and disbursement of funds; surveys for collection of technical data; formulation of schemes for technical education of women, handicapped and weaker sections of society; promotion of research and development; effective linkages with industry; technical teachers training and continuing education for technical personnel; norms and standards for courses, staff, physical infrastructure etc; approval for starting new technical institutions; and norms for granting autonomy to technical institutions. The Council sets up **National Board of Accreditation (NBA)** for assessment and accreditation of technical institutions. The Council arranges for inspections of institutions for their approval, increase in intake, starting of new institutions and change of location of institution.

Executive Committee (E.C.) of AICTE

The Council constitutes E.C with the following 20 members

- Chairman of the AICTE – Chairman
- Vice-Chairman of AICTE
- Member–Secretary of AICTE – (Member-Secretary)
- Ex-officio members (5)
  - (i) Secretary of Education, Government of India.
  - (ii) Member representing Ministry of Finance, Government of India.
  - (iii) Chairman, University Grants Commission.
  - (iv) Director, Institute of Applied Manpower Research.
- Nominated Members (12)
  - (i) Two Chairmen of Regional Committees of AICTE
  - (ii) Three Chairmen of Boards of Studies of AICTE
  - (iii) Four out of eight members of Council representing States and Union Territories
  - (iv) Four experts in relevant areas of technical education nominated by Chairman, AICTE.

The E.C. meets as and when required and observes rules of procedures in regard to the transaction of business as per the provisions of the Act.

The minutes of the meetings of E.C are placed before the Council for its approval.

Boards of Studies and Regional Committees

The AICTE ensures effective coordination and development of education through All India Boards of Vocational Education, Technician Education, Engineering and Technology (U.G and P.G), Management and Pharmacy. The Boards of Studies advise E.C. on academic matters in their areas of concern including, norms, standards, model curricula, structure of courses, model facilities, etc.

The administrative wing of AICTE is strengthened by establishment of Regional Committees at Kanpur (northern Region), Chennai (Southern Region), Mumbai (Western Region), Kolkata (Eastern Region), Bangalore (South-West Region), Bhopal (Central Region) and Chandigrah (North-West Region).

Each Regional Committee (R.C) is represented by Chairman, ex-officio and nominated members. All nominations are made by Chairman AICTE. The Regional Committee performs the duties in accordance with direction of the Council and provides necessary information to head office from time-to-time.

Rules and Regulations made by the Council shall be laid before each House of Parliament, while it is in session for a total period of thirty days.

CHAPTER 6

The University Grants Commission (U.G.C.) Act, 1956

Salient Features

Preamble

There are around five hundred pharmacy colleges in the country imparting Bachelor of Pharmacy (B.Pharm.) programme. The course is offered by university departments, constituent colleges, Government and private colleges affliated to various Indian Universities. The Universities offering various courses in higher education including B. Pharm. and M. Pharm. programmes are established under UGC Act, 1956.

It is an Act for establishment of a **University Grants Commission** for the purpose of effective co-ordination and determination of standards in Indian Universities.

University means a University established under a Central or State Act recognized by UGC in accordance with the regulations under this Act. On the advice of UGC, the Central Government may notify an institution for higher education other than a university as **"Deemed to be a university"** for the purpose of this Act.

The Commission is a body corporate having perpetual succession and a common seal, and shall by the said name sue or be sued. The head office of the Commission is at Delhi.

Composition of the Commission

It consists of a Chairman, a Vice-Chairman and ten other members appointed by the Central Government. The Chairman should not be from amongsts officers of the Central or State Governments.

Ten members nominated by the Central Government shall be as follows:

(i) Two Central Government officers.

(ii) Maximum four shall be from teachers of Universities.

(iii) The remainder shall be Vice-Chancellors of Universities or experts from engineering, medical, legal or any other profession, or persons with experience in agriculture, commerce, forestry or industry.

The Chairman holds office for a term of five years or until he/she attains the age of 65 years. The Vice-Chairman holds office for 3 years or until the age of 65 years. The term of all other members is for three years.

The re-appointment of Chairman, Vice-chairman and members is possible. The Commission meets at such times and places and transacts business in accordance with the regulations made under this Act. The Commission shall have its own fund and all sums which may, from time to time, be paid by the Central Government.

Powers and Functions of the Commission

The Commission, in consultation with the universities or other bodies concerned takes all such steps which are necessary for the promotion and co-ordination of higher education, determination and maintenance of standards of teaching, examinations and research in universities. The Commission is charged with following responsibilities.

(i) To enquire into financial needs of universities; allocate and disburse grants to universities, deemed to be universities and centers of excellence.

(ii) To advise any university on measures necessary for improvements of university education.

(iii) To advise Central and State Governments on allocation of any grants to the universities out of the Consolidated Fund of India or the State, as the case may be.

(iv) To advise any authority, if asked for, on the establishment of a new university and on expansion activity of existing university.

(v) To collect information on all matters related to university education in India.

(vi) To advise Central Government or any State Government or Union Territory on any question referred to it.

(vii) To perform other necessary functions for advancing the cause of higher education in India.

Section 12A of the Act deals with regulation of fees and prohibition of donation.

The university or the college is required to be approved by UGC under section 2 (f) and 12(B) for receiving grants.

The Commission prepares budget for the ensuing year showing the estimated receipts and expenditure and forwards it to the Central Government. The Annual Report is prepared by the Commission once in a year giving full account of its activities during previous year and the same is laid before both Houses of Parliament.

The Central Government may, by notification in the Official Gazette, makes Rules to the Act, while the Commission is empowered to make Regulations, consistent with this Act and Rules. The Commission shall be guided by the Central Government on policy matters relating to national purposes. In case of any dispute between Central Government and the Commission, the decision of the Central Government shall be final.

CHAPTER 7

The Drugs and Magic Remedies (Objectionable Advertisements) Act, 1954 and Rules, 1955

Ethical advertisement is the way of life in this competitive world of trade, and pharmaceuticals are no exception to this practice of advertisement. Advertisement of products is day-by-day increasing because of availability of impressive advertisement channels including TV, electronic media and proactive press. However, the cut-throat competition in advertisements and unethical approach to canvassing are required to be prohibited.

Indian population, especially in rural areas, even today is greatly influenced by the so-called miraculous power of magic remedies. The use of Talisman, Taits, Sacred Threads, Kavachas, different types of bone structures, Divine Bhasmas, Mantras and other so-called miraculous powers is commonly witnessed by large segment of society, even today. The practice of Bhanamati and related beliefs are predominantly prevailing in tribal areas. Any advertisements orally or in writing about the cure of diseases mentioned under Schedule to this Act are objectionable.

Objective

It is an Act to control the advertisement of drugs in certain cases and to prohibit the advertisement for certain purposes of remedies alleged to possess magic qualities. The Act extends to the whole of India except the State of Jammu and Kashmir. It was notified on 30th April, 1954 and amended in 1963.

Important Definitions

1. Drug

(i) It includes a medicine for internal or external use of human beings or animals,

(ii) Any substance intended to be used for or in the diagnosis, cure, mitigation, treatment or prevention of disease in human beings or animals,

(iii) Any article other than food intended to effect or influence in any way the structure or any organic function of the body of human beings or animals,

(iv) Any article intended for use as a component of any medicine, substance or article referred to in (i), (ii) and (iii) as above.

2. Magic Remedy

It includes Talisman, Mantras, Kavachas and any other charm of any kind which is alleged to possess miraculous power for or in the diagnosis, cure, mitigation, treatment or prevention of any disease in human beings or animals or for affecting or influencing in any way the structure or any organic function of the body of human beings or animals.

3. Advertisement

It includes any notice, circular, label, wrapper or other document and any announcement made orally or by any means of producing or transmitting light, sound or smoke.

4. Registered Medical Practitioner

(i) A person who holds a qualification granted by an authority specified in or notified under section 3 of the Indian Medical Degrees Act, 1916 or specified in the Schedules to the Indian Medical Council Act, 1956; *or*

(ii) A person who is entitled to be registered as a medical practitioner under any law for the time being in force in any State to which this Act extends relating to the registration of medical practitioners.

Advertisements Banned or Prohibited

1. Advertisement for procurement or miscarriage in woman or prevention of conception in woman; or

2. Maintenance or improvement of capacity of human beings for sexual pleasure; or
3. The correction of menstrual disorder in woman; or
4. The diagnosis, cure, mitigation, treatment or prevention of any disease, disorder or condition specified in the Schedule to this Act. The list of such diseases is finalized in consultation with Drugs Technical Advisory Board (DTAB) constituted under Drugs and Cosmetics Act, 1940.

Prohibition of Participation in Certain Advertisements

The participation of a person in publication of following advertisements is prohibited by the Act.

(i) An advertisement which directly or indirectly gives a false impression regarding true character of the drug; *or*

(ii) An advertisement which makes a false claim of the drug; *or*

(iii) An advertisement which otherwise is false or misleading in any material particular.

Penalties

The penalty for first conviction is imprisonment for 6 months or fine or both. For subsequent conviction, the penalty is imprisonment for one year or fine or both.

The Gazetted Officer authorized by State Government can enter and search at all reasonable times with assistants, if any, where he has reason to believe that an offence to violation of this Act is taking place, seize any advertisement which he/she believes is in contravention with the provisions of the Act and report such a seizure to the magistrate and take his/her orders as to the custody thereof. The Gazetted Officer can examine any record, register, document or any other objectionable material.

If the offence is committed by a company or institution, all the directors of the company or the persons present at the site of offence are liable to be prosecuted.

Nothing of this Act applies to a person who proves his innocence in the court or who proves his genuine efforts to stop such an offence or if he proves his good intentions to safeguard the provisions of this Act.

Advertisements Permitted

The following categories of advertisements are permitted under this Act.

1. Any signboard or notice displayed by registered medical practitioner on his premises indicating that treatment for any disease, disorder or condition specified in section three of the Schedule is undertaken in those premises.
2. Any book or treatises dealing with any of the matters specified in section three of the Schedule from a bonafide scientific or social stand point.
3. Any advertisement relating to any drug sent confidentially in prescribed manner only to a registered medical practitioner.
4. Any advertisement relating to any drug printed by the Government.
5. Any advertisement relating to a drug printed or published by a person for which permission was granted before the commencement of this Act.

The leaflets or literature accompanying packaging of drugs, advertisement of drugs in medical, pharmaceutical and scientific journals are permitted on conditions that the advertisement should contain only such information as is required for guidance of registered medical practitioners in respect of therapeutic indications of the drug, it's administration, dosage and side effects and the precautions to be observed in treatment with the drug. It shall be the responsibility of the advertiser to prove that any claim made in the advertisement with respect to the drug is true and not misleading or exaggerated.

The diseases, disorders or conditions under the Schedule for which uncontrolled and **misleading advertisements are prohibited** are appendicitis; blindness; blood poisoning; cancer; cataract; deafness; diabetes; diseases and disorders of brain, optical system, uterus; disorders of menstrual flow, nervous system, prostatic gland; epilepsy; fever (in general); female diseases (in general); fits; form and structure of female bust; stones of gall bladder, kidney and urinary bladder; gangrene; glaucoma; goitre; heart diseases and high or low blood pressure, hydrocele; hysteria; leprosy, leucoderma; obesity; paralysis; plague; pneumonia; rheumatism; sexual impotence; smallpox; stature of persons; sterility in women; tuberculosis; tumours; typhoid fever; ulcers of G.I. tract; venereal diseases including syphilis, gonorrhea, soft chancre, venereal granuloma and lympho-granuloma; and some other ailments or diseases.

Import or Export of Banned Advertisement

The involvement of a person in import or export activities pertaining to documents of objectional advertisements is punishable under this Act. The Central Government through the customs officer controls the import or export of prohibited materials under this Act. The custom collector shall retain the consignment of such material and dispose it of in accordance with the provision of Sea Custom Act, 1878. An aggrieved party can appeal to him within one week of the date of receipt of order of the custom collector and give him an undertaking in writing that consignment shall not be disposed without his approval. The customs collector shall pass an order to that effect. However, if ordered by the customs collector, the objectionable consignment is required to be returned to him within ten days.

Chapter 8

The Drugs and Cosmetics Act (DCA), 1940 and Rules, 1945

Drug is an essential commodity and is required to be regulated in terms of its import, manufacture, sale and distribution. The Central Government and State Government are charged with the responsibility of providing the drugs of desired quality to the needy patients and in order to ensure this primary obligation of the Government, the network is required to be developed to root out adulterated, misbranded and spurious drugs from the society.

The origin of DCA has a very interesting historical background. The enactment of the Act was an outcome of a sustained struggle by the medical and pharmaceutical fraternity of the country in pre-independent era against the British misrule in providing poor health care system to Indian citizens. The observation made by Drugs Enquiry Committee, representations made by Indian Medical Association, reports appeared in Indian Medical Gazette during 1920-30 and adverse press reports compelled the British rulers to initiate remedial measures for overcoming lacunae existing in the health care system. The Import of Drugs Bill introduced in 1937 was withdrawn due to heavy criticism and later on the Drugs Bill was introduced and the Drugs and Cosmetics Act, 1940 was enacted. The Rules were framed under this Act in 1945.

The Act is extended to the whole of India, including State of Jammu and Kashmir. It is divided into five chapters.

Chapter I	:	Introductory.
Chapter II	:	Drugs Technical Advisory Board (DTAB), Central Drugs Laboratory (CDL) and Drugs Consultative Committee.
Chapter III	:	Import of Drugs and Cosmetics.
Chapter IV	:	Manufacture, Sale and Distribution of Drugs and Cosmetics.
Chapter IV (A)	:	Provisions Relating to Ayurvedic, Siddha and Unani Drugs.
Chapter V	:	Miscellaneous.

There are two Schedules to the Act and 35 Schedules to the Rules framed under the Act.

It was in the year 1982, Schedules **E**, **I** and **L** were dropped, Schedules **G** and **H** were revised and Schedule **X** was introduced. In 1988, Schedule **M** incorporating **GMP (Good Manufacturing Practices)** was amended and Schedule **Y** Pertaining to clinical trials of newer drug formulations was incorporated. Schedules **T** of Ayurvedic Siddha and Unani medicines was introduced in 2000.

Important Definitions

1. Drug

It Includes:

(i) All medicines for internal or external use of human beings or animals and substances used for or in the diagnosis, treatment, mitigation or prevention of any disease or disorder in human beings or animals, including preparations applied on human body for the purpose of repelling insects like mosquitoes.

(ii) The substances other than food which may affect the structure or any function of the human body or used for the destruction of insects or vermin, which cause disease in human beings or animals, as specified from time to time by the Central Government by notification in the Official Gazette.

(iii) The substances used as components of a drug, including empty gelatin capsules.

(iv) The devices used for internal or external use in the diagnosis, treatment, mitigation or prevention of disease or disorder in human beings or animals, as may be specified from time to time, by the Central Government by notification in the Official Gazette after consultation with the Drugs Technical Advisory Board (DTAB).

2. Cosmetic

It means any article intended to be sprayed, poured, rubbed or sprinkled on, or introduced into, or applied to the human body or its any part for cleansing, beautifying, promoting attractiveness or altering the appearance. It also includes any articles intended for use as a component of cosmetic.

3. Ayurvedic, Siddha or Unani Drugs

These include all medicines used for internal or external purposes or used in the diagnosis, treatment, mitigation or prevention of disease or disorder in human beings or animals and manufactured exclusively in accordance with the formulae described in the authoritative books of Ayurvedic, Siddha and Unani Tibby Systems of Medicines specified in the First Schedule to the Drugs and Cosmetics Act, 1940.

4. Gudakhu

It is a tobacco product used for rubbing against human teeth. It contains tobacco powder, lime and molasses alongwith red mineral matter. It is a cosmetic within the provisions of the Act.

5. Patent or Proprietary Medicine

It means:

(i) In relation to Ayurvedic, Siddha or Unani System of Medicine, all formulations containing only such ingredients mentioned in the formulae described in the authoritative books of Ayurvedic, Siddha or Unani System of Medicine specified in First Schedule to the Act but does not include the medicine administered by parenteral route.

(ii) In relation to any other system of medicine including, allopathic, a drug presented in a form ready for internal or external administration of human beings or animals and which is not included for the time being in the editions of Indian Pharmacopoeia or any other Pharmacopoeia.

6. Misbranded Drug

A drug is considered as a misbranded drug:

(i) if it is not labeled in the prescribed manner,

or

(ii) if it is so colored, coated, powdered or polished that damage is concealed or it is made to appear of better or greater therapeutic value than it really is,

or

(iii) if the label or container or anything accompanying the drug bears any statement, design or device, which makes any false claim for the drug or gives misleading information.

7. Adulterated Drug

A drug is considered to be adulterated:

(i) if it consists in whole or in part of any filthy, putrid, or decomposed substance

or

(ii) if it has been prepared, packed or stored under poor sanitary conditions whereby, it may have been contaminated with filth and rendered injurious to health

or

(iii) if container of the drug is composed in whole or in part of any poisonous substance which may render the contents injurious to health

or

(iv) if it contains a color other than one which is prescribed

or

(v) if the drug contains any harmful or toxic substance which may render it injurious to health

or

(vi) if the drug is admixed with any substance so as to reduce its quality or strength.

8. Manufacture in relation to Drug or Cosmetic

Any process fully or partly used for making, altering, ornamenting, finishing, packing, labeling, breaking up or otherwise treating or adopting any drug/cosmetic with a view to its sale or distribution but does not include the compounding or dispensing of any drug or the packing of any drug or cosmetic in the ordinary course of retail business.

9. Spurious Drug

A drug is deemed to be a spurious drug:

(i) if it is imported under a name which belongs to another drug

or

(ii) if it is an imitation of or a substitute for another drug or if it resembles another drug in a manner likely to deceive or bears upon it or its label or container the name of another drug

or

(iii) if it has been substituted wholly or in part by another drug substance

or

(iv) if it claims to be the product of a manufacturer or company of whom it is not truly a product.

10. Misbranded Cosmetic

A cosmetic shall be deemed to be misbranded:

(i) if it contains a color which is not prescribed

or

(ii) if it is not labeled in prescribed manner

or

(iii) if the label or container or anything accompanying the cosmetic bears any statement which is false or misleading.

11. Spurious Cosmetic

A cosmetic shall be deemed to be spurious:

(i) if it is imported under a name which belongs to another cosmetic

or

(ii) if it is an imitation of or a substitute for another cosmetic; resembles another cosmetic in a manner likely to deceive; or bears upon it or upon its label or container the name of another cosmetic

or

(iii) if the label or container bears the name of an individual or a company purporting to be the manufacturer of the cosmetic which individual or company is fictitious or does not exist

or

(iv) if it purports to be the product of a manufacturer of whom it is not truly a product.

Drugs Technical Advisory Board (DTAB)

It is a Statutory Board constituted by the Central Government under the provision of this Act to advise the Central Government and State Governments on all the technical matters related to the Act and also to set the guidelines for types of formulations as and when asked for by the Central Government. It is a technical advisory body represented by the Ex-officio, nominated and elected members. The total strength of DTAB is eighteen representing different facets of pharmacy and medical profession in the country. The Director General of Medical and Health Services, Government of India, is the Chairman of DTAB, while Drugs Controller General of India is the Member Secretary. The Head Office of DTAB is at Ministry of Health and Family Welfare, Government of India, Nirman Bhavan, New Delhi.

Composition of DTAB

The composition of Drugs Technical Advisory Board is shown in Table 8.1

Table 8.1

Ex-officio Members (8)	Elected Members (5)	Nominated Members (5)
(i) Director General of Medical and Health Services, Government of India (Chairman).	One Professor in pharmaceutics or pharmaceutical chemistry or pharmacognosy elected by the Executive Committee of PCI from any University or affiliated pharmacy college.	Two members to be nominated by Central Government who are incharge of Drugs Control Department of the State or Union Territory.
(ii) Drugs-Controller General of India (Member Secretary).	One professor in medicine or therapeutics in any of the Government or affiliated medical College elected by Executive Committee of Medical Executive Committee of Medical Council of India.	Two Government Analysts to be nominated by Central Government anywhere from the country.
(iii) Director, Central Drug Research Institute (CDRI) Lucknow, U.P	One professor in pharmacology and toxicology to be elected by the Governing Body of Indian Council of Medical Research (ICMR)	One industrialist representing pharmaceutical industry to be nominated by Central Government
(iv) Director, Central Drug Laboratory, Kolkata	One member to be elected by the Central Council of Indian Pharmaceutical Association (IPA)	
(v) Director, Indian Veterinary Research Institute (IVRI), Izzatnagar, U.P	One member to be elected by the Indian Medical Association (IMA)	
(vi) Director, Central Research Institute (CRI), Kasauli, H.P.		
(vii) President, Pharmacy Council of India (PCI)		
(viii) President, Medical Council of India (MCI)		

The term of elected and nominated members is for three years. The Ex-officio members hold the office so long as they are in that specific position. The DTAB can form the sub-committees and also co-opt member experts for specific assignment even though they are not the members of DTAB. The DTAB takes the policy decisions pertaining to technical aspects of the Drugs and Cosmetics Act and Rules and sends the recommendations to the Ministry of Health and Family Welfare, Government of India, for its approval. DTAB ordinarily meets twice a year. For certain urgent matters, DTAB can be summoned with one week's notice. Sometimes very urgent matters can be, on priority basis, decided by the Ministry of Health and Family Welfare, Government of India. However, such decisions of the government are required to be ratified by DTAB within six months.

Drugs Consultative Committee (DCC)

It is the Advisory Body nominated by the Central Government for advising the Central and State Governments, as well as the DTAB on the matters pertaining to the uniform implementation of the provisions of DCA and Rules. The composition of DCC is represented by two representatives nominated by the Central Government and one representative each of the State Government and Union Territory. Usually, the Director of Drugs Control Administration or Drug Controller of State is nominated by the State Government or Union Territory on this Council.

Central Drugs Laboratory (CDL)

The Central Drugs Laboratory is headed by the Director with its headquarters at Kolkata. It is the Statutory Analytical Laboratory for drugs and cosmetics under DCA whose decision with regards to analysis is final in the court of law. The CDL is charged with the following responsibilities.

1. It takes up the analysis of samples of drugs and cosmetics sent by custom collectors and different courts.
2. As directed by the Central Government, it advises the Central Government and the State Governments and Union Territories on the matters pertaining to the analysis of drugs and cosmetics and also takes up analytical work of specific nature for samples sent by Central Government and State Governments.

3. It may take up the samples for analysis on payment of necessary fee for private parties, consumer organization, etc.
4. It is engaged in the research for the development of newer techniques of analysis of drugs and cosmetics.

For the analysis of following items, the Director of concerned laboratory as indicated, has all the powers of the Director of CDL under the provisions of the Act.

1. Director, Central Research Institute (CRI), Kasauli, H.P: for biological preparations such as vaccines, sera, toxins, toxoids, etc., and also bacteriophages, surgical sutures and ligatures.
2. Director, Indian Venterinary Research Institute (IVRI), Izzatnagar, U.P: for all biological products and other veterinary products meant for animals.
3. Director, Central Indian Pharmacopoeia Laboratory (CIPL) Ghaziabad, U.P: for all homoeopathic medicines and condoms.
4. Director, National Institute of Communicable Diseases:- Oral Polio Vaccine.

The sample for analysis is required to be sent in sealed cover by registered post addressed to Director of the laboratory concerned. Along with this a memorandum filled in by the person sending the sample for analysis in accordance with the procedure laid down should be accompanied separately in the same registered post. Another copy of this memorandum is required to be sent separately with the impression of the seal by registered post to the Director. The registered post of sample and memorandum is received by the officer deputed on behalf of Director or Director himself. The impression of the seal in both the cases is compared and its genuinity is confirmed and only then, the seal is opened. The sample is required to be kept under the custody of the Director or the officer identified by him until the analysis is over and thereafter, for a period of one year. On completion of the analysis, the protocol followed for analysis, analytical results and other relevant information are sent by registered post to the party concerned. Depending upon the urgency of the matter, priority for analysis is fixed. The decision given by CDL regarding analysis is final and not challengable in the court of law.

Wings of DCA	
1. Advisory Wing	(i) Drugs Technical Advisory Board (ii) Drugs Consultative Committee
2. Analytical Wing	(i) Central Drugs Laboratory (ii) Government Analysts (iii) Drug Testing Laboratories of the States
3. Administrative Wing	(i) Drugs Controller General of India (ii) Drugs Control and Licensing Authorities of States (iii) Drug Inspectors of Central and State Governments.

Schedules to the Act and Rules

There are two Schedules to the Act and 35 Schedules to the Rules.

The Schedules to the Act

1. ***First Schedule:*** It comprises the list of books of references for Ayurvedic, Siddha and Unani medicines. There are 57 books of Ayurveda, thirty books of Siddha and thirteen of Unani Tibb systems listed in the Schedule which are used for different formulations in accordance with the provisions of the Act.
2. ***Second Schedule*** : It comprises the standards to be complied with for imported drugs, manufacture of drugs, their sale, stocking and storage etc.

The Schedules to the Rules

Schedule A : Different forms for application to procure licence, renewal of licence, and for all other activities.

Schedule B : Rates of fees charged for analysis by CDL or State Drugs Laboratories.

Schedule C	: List of biological and other special products governed by special provisions.
Schedule C (I)	: List of other special products governed by special provisions.
Schedule D	: Class of drugs: extent and conditions of exemption
Schedule D (I)	: Undertaking of the manufacturer or his authorized agent required to be submitted along with application form for obtaining a registration certificate.
Schedule D (II)	: Undertaking of the manufacturer or his authorized agent required to be submitted along with application form for registration of a bulk drug or its formulation or its import into India.
Schedule E (I)	: List of poisonous substance under Ayurvedic, Siddha and Unani medicines.
Schedule F	: Requirement for operation of blood bank and / or preparation of blood components.
Schedule F (I)	: Provisions for bacterial vaccines, viral vaccines, antisera, diagnostic antigens, etc.
Schedule F (II)	: Standards for surgical dressings.
Schedule F (III)	: Standards for umbilical tapes.
Schedule FF	: Standards for ophthalmic preparations.
Schedule G	: Drugs required to be taken under medical supervision.
Schedule H	: List of prescription drugs.
Schedule J	: List of diseases or ailments which a drug may not purport to prevent or cure.
Schedule K	: Drugs exempted from certain provisions related to manufacturer.
Schedule M	: **GMP (Good Manufacturing Practices)** comprising requirements of factory premises, plant and equipment.

Schedule M-1 : Homoeopathic preparations requirements of factory premises, plants and equipments.

Schedules M-(II) : Cosmetics – requirements of factory premises for manufacture.

Schedules M-(III): Requirements of factory premises for manufacture of medical devices.

Schedule N : List of minimum equipment for running a pharmacy.

Schedule O : Standards for disinfectant fluids.

Schedule P : Life period of drugs.

Schedule P-1 : Pack sizes of drugs.

Schedule Q : List of colors, dyes and pigments permitted in cosmetics and soaps, list of colors permitted in soaps.

Schedule R : Standards for condoms of rubber latex and other mechanical contraceptives.

Schedule R-I : Standards for medical devices.

Schedule S : Standards for Cosmetics

Schedule T : GMP (Good Manufacturing Practices) for manufacture of Ayurvedic, Siddha and Unani medicines, G.M.P., machinery, equipment minimum manufacturing premises, etc.

Schedule U : Particulars required to be shown in manufacturing records; raw material and analytical records.

Schedule U (1) : Particulars required to be shown in manufacturing records.

Schedule V : Standards for patent or proprietary medicines.

Schedule X : Psychotropic substances.

Schedule Y : Requirements and guidelines on clinical trials for import and manufacture of new drug.

Different **Forms** used for a variety of operations pertaining to Drugs and Cosmetics are covered under **Schedule A** to Rules of DCA. They include:

Form 1 - Memorandum to the Central Drugs Laboratory (CDL).

Form 2 - Certificate of CDL for test or analysis.

Form 8 - Application for licence to import drugs except, schedule X (Roman X) and Licence for this is issues in Form ten.

Form 8-A - Application for licence to import schedule X drugs and licence for this is issued in **Form 10-A.**

Form 9 - Undertaking accompanying application for import licence.

Form 11 - Licence for import of drugs of examination, test or analysis, while the application for this licence is made in **Form 12.**

Form 11-A - Licence to import drugs by a Government hospital or autonomous medical institution, while the application for this is made in **Form 12-AA.**

Form 12-A - Application for permit to import small quantities of drugs for personal use, while permit is issued in **Form 12-B.**

Form 13 - Certificate of test or analysis by Government Analyst under section 25 (1) of DCA, 1940.

Form 13-A - Certificate of test or analysis by Government Analyst under Section 33 H of DCA, 1940.

Form 14-A - Application for test or analysis of a drug from purchaser under section 26 of DCA, 1940 and the certificate for the same is issued in **Form 14-B.**

Form 15 - Order requiring a person not to dispose of stock in his possession.

Form 16 - Receipt for seized stock of drugs or cosmetics, records, register, documents or material.

Form 17 - Intimation to person for taking sample.

Form 17-A - Receipt for samples of drugs or cosmetics taken where fair price is tendered.

Form 18 & 18-A - Memorandum to Government Analyst.

Form 19 - Application for grant or renewal of a licence to sell, stock, exhibit or distribute drugs other than schedule X.

Form 19-A - Application for grant or renewal of a **restricted licence** to sell, stock, exhibit or distribute drugs by retail by dealer who does not engage qualified person. The licence for this is issued in **Form 20-A.**

Form 19-AA - Application for grant or renewal of licence to sell, stock, exhibit or distribute drugs from a motor vehicle. The licence for this is issued in **Form 20-BB.**

Form 19-B - Application for a licence to sell, stock, exhibit, sale or distribute Homoeopathic medicine. The licence for retail is issued in **Form 20-C**, while the licence for wholesale is given in **Form 20-D**. The certificate for renewal of licence is issued in **Form 20- E**.

Form 19-C - Application for grant or renewal of a licence to sell, stock, exhibit or distribute schedule X drugs. The licence for retail sale is issued in **Form 20-F,** while the licence for the same by wholesale is given in **Form 20-G.**

Form 20 - Licence to sell, stock, exhibit or distribute drugs other than specified in schedules C, C (1) and X by retail.

Form 20-B - Licence to sell, stock, exhibit or distribute by wholsale plain drugs other than those specified in schedules C, C (1) and X.

Form 21 - Licence to sell, stock, or exhibit or distribute by retail Schedules C and C (1) drugs.

Form 21-A - Restricted licence to sell, stock, or exhibit or distribute by retail schedule C (1) drugs for dealers not engaging services of qualified person.

Form 21-B - Wholesale licence for sell, stock, exhibit or distribute Schedules C and C (1) drugs.

Form 21-BB - Licence to sell by wholesale or to distribute drugs specified in Schedules C and C (1) from a motor vehicle.

Form 21-C - Certificate of renewal of licence to sell, stock, exhibit or distribute drugs.

Form 21-CC - Certificate of renewal of licence to sell, stock, exhibit or distribute drugs by wholesale from a motor vehicle.

Form 24 - Application for grant or renewal of manufacturing licence for drugs other than Schedules C, C (1) and X drug. The licence for manufacture of this category is issued in **Form 25.**

Form 24-A - Application for grant or renewal of a loan licence to manufacture drugs other than Schedules C, C (1) and X drug. The loan licence for manufacture of this category is issued in **Form 25-A.**

Form 24-B - Application for grant or renewal of repack licence for sale for drugs other than Schedules C, C (1) and X drugs. The licence for repack of this category is issued in **Form 25-B.** The certificate of renewal of such a repack licence for sale is issued in **Form 26-B.**

Form 24-C - Application of grant or renewal of a licence for manufacture of homoeopathic medicines or potentised preparations from back potentcies by licensees holding licence in **Form 20-C.** The licence to manufacture homoeopathic medicines is issued in **Form 25-C.** The certificate of renewal of such a licence is in **Form 26-C.**

Form 24-D - Application for grant/renewal of licence for manufacture for sale of Ayurvedic/Siddha or Unani drugs. The licence for manufacture of this category is issued in **Form 25-D.** The certificate of renewal licence for manufacture is issued in **Form 26-D.**

Form 24-E - Application4 for grant/renewal of a loan licence to manufacture for sale Ayurvedic /Siddha or Unani drugs. The loan licence for manufacture of this category is issed in **Form 25-E.** The certificate of renewal of loan licence to manufacture such drugs is issued in **Form 26-E.**

Form 24-F - Application for grant/renewal of manufacturing licence of drugs in Schedule X and not in Schedules C and C (1). The licence for manufacture of this category is issued in **Form 25-F.**

Form 26 - Certificate of renewal of licence to manufacture other than schedule X drugs, while certificate of renewal of loan licence to manufacture such drugs is issued in **Form 26-A.**

Form 26-E-1 - Certificate of Good Manufacturing Practices (GMP) to manufacture of Ayurveda, Siddha or Unani drugs.

Form 26-F - Certificate of renewal of licence to manufacture for sale of Schedule X drugs.

Form 26-G - Certificate of renewal of licence for operation of Blood Bank for processing human blood or its components.

Form 26-H - Certificate of renewal of licence to manufacture large volume parenterals, sera and vaccines of Schedules C and C (1) categories.

Form 26-I - Certificate of renewal of licence for manufacture of blood products.

Form 27 - Application for grant/renewal of a manufacturing licence of drugs in Schedules C and C (1), excluding those specified in Part X B and schedule X. The licence for manufacture for this category is issued in **Form 28.**

Form 27-A - Application for grant/renewal of a loan licence to manufacture Schedules C and C (1) drugs, excluding those specified in Part X B and schedule X. The licence for manufacture of this category of drugs is issued in **Form 28-A.**

Form 27-B - Application for grant/renewal of manufacturing licence for Schedules C, C (1) and X. The licence for manufacture of this category is issued in **Form 28-B.**

Form 27-C - Application for grant/renewal of a licence for Operation of Blood Bank for processing of whole blood and / or preparation of blood components. The licence for this is issued in **Form 28-C.**

Form 27-D - Application for grant or renewal of a manufacturing licence of large volume parenterals/sera and vaccines excluding Schedule X drugs.

Form 27-E - Application for grant/renewal of manufacturing licence of blood products.

Form 28-D - Licence to manufacture for sale or distribution of large volume parenterals, sera and vaccines specified in Schedules C and C (1) excluding those specified in Schedule X.

Form 28-E - Licence to manufacture and store blood product for sale or distribution.

Form 29 - Licence to manufacture drugs for examination, test or analysis. Application for the same is made in **Form 30.**

Form 31 - Application for grant/renewal of a licence to manufacture cosmetics. The licence for manufacture of cosmetics is issued in **Form 32**, **Form 31-A** - Application for loan licence to manufacture cosmetics, the loan licence for manufacture of cosmetics is issued in **Form 32-A.**

Form 33 - Certificate of renewal of licence to manufacture cosmetics.

Form 33-A - Certificate of renewal of loan licence to manufacture cosmetics.

Form 34 - Certificate of test or analysis of cosmetic by CDL or Government Analyst.

Form 35 - Form for maintaining Inspection Book.

Form 36 - Application for grant/renewal for carrying out tests on drugs/cosmetics or raw materials used in manufacture thereof on behalf of licensee for manufacture for sale of drugs/cosmetics. Approval for the same is issued in **Form 37,** whereas Certificate of Renewal for the same is issued in **Form 38.**

Form 39 - Report of test or analysis by approved institution.

Form 40 - Application for issue of Registration Certificate for import of drugs into India. The Registration Certificate is issued in **Form 41.**

Form 44 - Application for grant of permission to import or manufacture a New Drug or to undertake clinical trial.

Form 45 - Permission to import finished formulation of a new drug.

Form 45-A - Permission to import new bulk drug substance (raw material).

Form 46 - Permission/Approval for manufacture of new drug formulation.

Form 46-A - Permission/Approval for manufacture of new bulk drug substance (raw-material).

Form 47 - Application for grant or renewal of approval for carrying out tests on Ayurvedic, Siddha and Unani drugs or raw material used in manufacture thereof on behalf of Licensee for sale of Unani, Ayurveda and Siddha drugs. The approval for test or analysis of this category is issued in **Form 48** and certificate of renewal for carrying out tests or analysis is issued in **Form 49.**

Form 50 - Report of test or analysis by approved laboratory.

Licensing and Controlling Authorities

Through the ninth amendment to the Drug and Cosmetic Rules in 1989, for the first time, the requisite qualification separately for Licensing Authority and Controlling Authority are indicated under 49 A and 50 A respectively. The qualification required for both Licensing Authority and Controlling Authority is a graduate in pharmacy or pharmaceutical chemistry or in medicine with clinical pharmacology or microbiology from university established in India and a minimum of five years in manufacturing or testing of drugs or enforcement of the provisions of the Act. The qualifications are, however, applicable from the date of amendment and not for those Inspectors and Government Analysts who were holding those positions on 12 - 4 - 1989.

Inspector

The Drug Inspectors are appointed both by State Government and Central Government for specific areas or for specific category of activity. There could be a set of inspectors exclusively for the manufacture of drug formulations. Drug Inspector works under the Drug Controling Authority under State Government or Central Government as the case may be. An inspector has been charged with the responsibility of ensuring strict implementation of Drugs and Cosmetics Act in the area of his/her jurisdiction.

1. Qualification for Drug Inspector

The person should not have direct or indirect financial interest in any of the activities concerned with import, manufacture, sale or distribution of drugs.

A graduate in pharmacy or pharmaceutical sciences or medicine with specialization in clinical pharmacology or microbiology from a University established in India, is eligible for the post of Inspector.

For the purpose of Schedules C and C (1) drugs, (i) a drug inspector with at least eighteen months of experience in manufacturing of at least one substance specified in Schedules C and C (1) or (ii) a drug inspector with at least three years experience in inspecting the firms manufacturing Schedules C and C (1) drugs or (iii) a drug inspector with minimum of eighteen months experience in testing of at least one of the substances in schedules C and C (1) in a laboratory approved for the purpose.

The requirement of these qualifications shall not, however, apply to those persons appointed as Inspector on or before 18th October, 1993.

Every Inspector shall be deemed to be a public servant under Section 21 of the Indian Penal Code.

2. Duties of Inspectors of Premises Licensed for Sale

Subject to instructions of Controlling Authority, it shall be the duty of an Inspector authorized to inspect premises licensed for the sale of drugs,

(i) To inspect, not less than once a year, all establishment for sale.

(ii) To satisfy himself that conditions of licences are being observed.

(iii) To procure and send the drug for test or analysis if he has reason to suspect that drug is sold or stocked in contravention of the provisions of the Act or Rules.

(iv) To investigate complaint in writing.

(v) To maintain a record of inspections.

(vi) To make necessary enquiries.

(vii) To institute prosecutions in respect of breaches of the Act and Rules.

(viii) When authorized by the State Government, to detain imported packages which he has reason to believe contain drugs, the import of which is prohibited.

3. Duties of Inspectors Specially Authorized to Inspect Manufacture of Drugs or Cosmetics

Subject to instructions of Controlling Authority, the following duties are performed:

(i) To inspect, not less than once a year, all premises licenced for manufacture of drugs or cosmetics.

(ii) To satisfy himself that conditions of licence are fulfilled.

(iii) To inspect plant, process of manufacture and standardization, storage, technical qualifications and other details for Schedules C and C (1) drugs.

(iv) To send detailed inspection report to Controlling Authority.

(v) To take samples for test or analysis in accordance with Rules.

The Inspector, except for official business or when required by law, shall not disclose any information acquired by him.

If he has sufficient reasons to believe that violation of provisions of DCA is taking place, he/she may and seize records or ask the manufacturer not to sell the drugs for a period of 20 days.

Drug Inspector if required may take the xerox copies of the seized documents signed by the owner of the documents.

Drug Inspector is supposed to carryout a routine inspection at least once in a year of a shop or a manufacturing unit within his area. The inspection should be generally carried out at reasonable time preferably during working hours. However, if he/she has sufficient reasons to believe that contravention of the provision of DCA is taking place, he/she accompanied by sufficient force may raid the premises and seize the documents, records or the medicines, as the case may be.

The sample withdrawn or seized from drug store should be divided into four parts. The seizure of medicine should be carried out in accordance with the code of criminal procedure, 1898 in presence of witnesses. The samples should be sealed and the seal of drug store owner should also be allowed. In case of injectables, four different ampoules of same batch are seized. The payment of fair price of seized material is made to the drugs store owner or in case of refusal for accepting money, the receipt is prepared by Drug Inspector separately and the form is filled up. Any action of seizure or raid is required to be informed to Judicial Magistrate of that area immediately. Out of four samples confiscated, one is retained by the Inspector, one is sent to Government analyst, one is given back to drugs store owner and fourth sample is sent to the manufacturer.

The sample sent by an Inspector to Government Analyst shall be by registered post or by hand in sealed packet enclosed with memorandum in **Form 18** in an outer cover addressed to the Government Analyst. A copy of memorandum and a specimen impression of seal is sent separately to the Government Analyst by registered post.

In case of raid or seizure of medicine at manufacturing unit, three seized samples are prepared following same procedure. One sample is retained by the Inspector for producing in court. The second sample is handed over to manufacturer and third sample is sent for analysis to Government Analyst.

After receipt of report of analysis, action is taken accordingly. If the report is satisfactory, regular sale is allowed. If it is not satisfactory, further legal action is taken up.

Any physical assault or a threat in writing or on telephone to Inspector while he is discharging his duties is considered as an offence punishable with imprisonment up to three years or fine or both.

Forms

Form 15 - It is an order given by Inspector requiring a person not to dispose of any stock in his possession.

Form 16 - It is the receipt the Inspector tenders for the seized material.

Form 17 - It is the intimation to the person from whom sample is taken.

Form 17-A - It is receipts of samples of drugs/cosmetics taken where fair price is tendered.

Form 18 - It is the memorandum to be sent by Drug Inspector to Government analyst for the purpose of analysis of seized formulation.

Government Analyst

The Government Analysts are appointed by Central Government and State Government for the purpose of test or analysis of drugs and cosmetics. They are employed in Central Drugs Laboratory and Drug Testing Laboratories of States and Union Territories.

Qualifications

The person appointed should have no financial interest in import, manufacturing, sale or distribution of drugs directly or indirectly. The person should be:

1. A graduate in medicine/science/pharmacy/pharmaceutical chemistry with a minimum of five years of experience after graduation in testing or analysis of drugs and pharmaceuticals in a laboratory under the control of Government Analyst/Head of the approved analytical institution.

or

2. a post-graduate in medicine/science/pharmacy/pharmaceutical chemistry/associateship diploma of institution of chemists, India with analysis of drugs and pharmaceutical as a special subject and at least three years of cxperience (in all cases) in testing or analysis of drugs and pharmaceuticals under the control of Government Analyst/Head of approved analytical institution.

For Schedules C and C (1) drugs, qualification as indicated above under clause (a) or (b) and having degree in medicine/physiology/pharmacology/microbiology/pharmacy should have experience in testing of said items in approved laboratory or institution for a minimum period of six months. For persons with other qualifications under clause 1. or 2. a minimum of three years experience of analysis in approved laboratory or institution is mandatory.

For veterinary biological products such as antisera, toxoids, vaccines, diagnostic antigens etc., the person appointed shall be a graduate in veterinary science/pharmacy/medicine/science with not less than five years experience in standardization of biological products.

or

Post-Graduate in veterinary science/medicine/general science/pharmacy/pharmaceutical chemistry with not less than three years experience in standardization of biological products.

Duties of Government Analyst

1. To analyze the samples sent by the Inspector, Custom Officer or other persons under provisions of Chapter IV of the Act and prepare a detailed report of analysis in triplicate. The complete protocol of test of analysis should be given. The report should be sent in sealed cover to custom Department or Drug Inspector as the case may be.
2. Forward the reports to the Government about work carried out, research undertaken, publications, if any and keep the information up-to-date in drugs and pharmaceuticals.

The Government Analyst should verify the seal before taking up analysis and ensure proper custody of the sample sent for analysis.

The analytical report should be supplied to the inspector or person concerned in triplicate in **Form 13.**

An application from a purchaser for test or analysis is made in **Form 14-A** and report of analysis is made available in **Form 14-B.**

Import of Drugs and Cosmetics

The import of drugs and cosmetics is regulated by the provisions of this Act.

Prohibition of Import of Certain Drugs and Cosmetics

The following categories of drugs and cosmetics are prohibited from import:

1. Drugs or cosmetics which are not of standard qualities.
2. Drugs or cosmetics which are misbranded, spurious and adulterated.
3. Drugs or cosmetics for import of which licence is required.
4. Any patent or proprietary medicine without true formula or list of active ingredients and their quantities.
5. Any drug or formulation which claims to prevent or cure diseases mentioned schedule J.
6. Any drug or cosmetic for which manufacture, sale or distribution is prohibited in country of its origin.
7. Any drug which is not packed or not labeled in conformity with the Rules of the Act.
8. Any cosmetic containing an ingredient which may render it unsafe or harmful.
9. Any drug or cosmetic the import of which is prohibited by Act.

Exemptions

1. The import of small quantities of any drug, subject to prescribed conditions, is permitted for test, analysis or personal use.
2. The Central Government, in consultation with DTAB, may permit import of any drug or class of drugs not being of standard quality.

The Commissioner of Customs or any officer authorized by the Central Government may detain any imported package of drug or cosmetic which he/she suspects to have been imported in contravention of the provisions of this Act and report the same to the Drugs Controller General of India and if required, send the sample for analysis to Central Drugs Laboratory.

The Central Government makes rules in consultation with DTAB for requirements of import licence, list of drugs for which licence is required, methods of analysis, the diseases for which imported drug may not claim cure, prescribe the places of import, conditions and all related matters.

Application and Duration of Import Licence and Registration Certificate

An application for import licence is made to licensing authority in **Form 8** for drugs excluding **Schedule X** and in **Form 8-A** for **schedule X** drugs. The licence is issued in **Form-10** or**10-A** as the case may be.

The application for registration certificate is made to Licensing Authority in **Form 40** and registration certificate is issued in **Form 41.**

The application for both import licence and registration certificate may be made by manufacturer himself or his authorized agent in India having a valid licence.

Both the import licence and registration certificate are valid for a period of *three* years from date of issue. If the application is made three months in advance before expiry of licence or certificate, it is valid until orders are passed on application.

Permitted Places for Import of Drugs

The import of drug into India is permitted only from following places:

1. By rail	:	(i)	Ferozepur Cantonment and Amritsar railway stations for drugs from Pakistan
		(ii)	Ranaghat, Bongaon and Mohiassan railway stations for drugs from Bangladesh
2. By road	:		Raxual for drugs from Nepal
3. By sea	:		Chennai, Kolkata, Mumbai, Nhava Sheva, Kandla, and Cochin
4. By air	:		Mumbai, Chennai, Kolkata, Delhi, Ahmedabad and Hyderabad.

Conditions of Import Licence

The importer has to fulfil the conditions that are stipulated in the Rules and also comply with following conditions.

1. The manufacturer shall observe undertaking given in **Form 9.**
2. The licensee should maintain a proper record of imported drug wherein the entries should be made serially for the stock of imported material, its distribution, persons to whom the imported drug is issued, price charged, remaining and quantity of imported drugs. The drug imported for the purpose of test or analysis or the new drugs imported are not for general use.
3. The importer should maintain all proper storage facilities for drugs imported as required in accordance with the provisions of the Act.
4. The importer should permit the inspector or officer on behalf of the State Government or Central Government without notice to inspect the premises, stocking facilities, records, analytical details, and sale of imported substance.
5. The licensee should withdraw the substance from market if asked to do so by authority, if found that the substance is substandard.

An import licence is for one category of drug from single manufacturer abroad, or it could be for more drugs from same manufacturer from one location. Separate licence is required for import of drugs from different manufacturers or from the same manufacturer located at different places.

Other Features of Import

- No new homoeopathic medicine can be imported without permission in writing from the Licensing Authority.
- Small quantities of a new drug may be permitted for import by a Government hospital or Autonomous medical institution for the treatment of patient suffering from life threatening disease, subject to fulfillment of conditions laid down for the purpose.

- Small quantities of drugs for examination, test or analysis may be imported subject to the conditions that the licensee shall use the drug exclusively, for the purpose for which it is imported; the licensee shall allow any inspector authorized by the licensing authority to inspect premises without prior notice and investigate the manner in which substances are being withdrawn and used. The licence is issued in **Form 11.** The licensee should maintain all the records and comply with conditions stipulated for licence.
- Small quantities of drugs for personal use are permitted to be imported, provided it is carriedout by the passenger in his/her baggage and the drug is declared to the Customs Authority. However, quantity only up to 100 average doses of such drug is permitted for import.

Suspension and Cancellation of Import Licence or Registration Certificate

If the manufacturer or licensee fails to comply with any conditions of the registration certificate, the licensing authority may after giving him an opportunity to show cause may suspend or cancel the registration certificate for such period as it thinks fit. However, the aggrieved person may appeal to the Central Government within 30 days against such order and the decision of the Government in this regard shall be final.

Offences and Penalties

1. An offence of any adulterated (section 9-A) or spurious drug (section 9-B) or cosmetic (section 9-D) being imported into the country in violation of provisions of the Act is punishable with imprisonment up to three years and a fine up to ₹ 5000.00 .
2. If any drug or cosmetic other than one referred to under (a) if illegally imported can attract punishment of six months imprisonment or fine of ₹ 500.00 or both.
3. Any drug or cosmetic imported in contravention of provisions of any notification issued under Section 10-A shall be punishable with imprisonment up to three years or fine up to ₹ 5000.00.

Schedule M

GMP (Good Manufacturing Practices) and Requirements of Premises, Plant and Equipment

In order to ensure production of quality drug formulation, it is necessary on the part of the manufacturer to follow well-established and ethical approach involving different operations of manufacture. It was on several occasions discussed in professional meetings and conferences that there is a need for well-set mandatary guidelines required to be followed by manufacturers of different dosage formulations. It was with this background, Good Manufacturing Practices under Schedule M were made mandatory conditions for manufacturin g operations of pharmaceutical formulations.

The quality of drug formulations is the sole responsibility of the manufacturer. He has to ensure the production of desired quality formulations and their stability until the formulation reaches the consumer across the retailing counter. Schedule M is covered under Rules 71, 74, 76 and 78 and is in two parts.

Part I deals with GMP relating to factory premises and materials.

Part II deals with requirement of plant and equipment.

PART I

(Salient Features) Factory Premises and Materials

- **General Requirements**

  Good location; free from contamination due to sewage, drain, fumes, dust, smoke, etc.; hygienic conditions; prevention of entry of insect/rodents; interior surface of premises should be smooth; adequate lighting; proper ventilation; humidity control; underground drainage; concealed electrical and sanitary fittings in the premises; supply of pure water; regular cleaning and disinfection of premises; proper treatment of waste water; and pollution control and disposal of pollutants.

- **Warehousing Area**

  Adequate area for orderly warehousing of various categories of materials; adapted to ensure good storage condition; protection from adverse weather conditions; separate earmarked areas in same warehouse for quarantine status; separate sampling area; segregation for storage of rejected, recalled or returned materials; safe and secure areas for NDPS and hazardous substances; safe storage of printed packaging material; separate dispensing areas for Beta lactum, sex hormones, cytotoxic substances and other special categories; and regular checks and rodent control.

- **Sterile Products**

  Separate enclosed area with air locks; air supply through HEPA filters; routine microbial counts; laminar flow cabinets availability and access restricted only to authorized persons.

- **Working Space**

  Adequate space for orderly placement of equipment and material; and separate storage area for raw material "under test", "approved" and "rejected". The pipe-work, electrical fittings and ventilation openings should be properly designed.

- **Health, Clothing and Sanitation of Workers**

  The workers should be free from contagious diseases. It covers regular medical check-up facilities; proper toilet facility at a distance; and personal cupboards and change room for workers.

- **Medical Services**

  First-aid facility; medical examination of workers and all other staff at the time of recruitment; periodic medical check-up of all staff members once in a year; availability of services of physicians at short notice; proper facilities for vaccinations; etc.

- **Sanitation in Manufacturing Premises**

  No accumulated waste; no dust particles as far as possible; proper disinfection and cleaning of premises; and no stagnant water. The manufacturing premises should be used for specific purpose for which it is designed.

- **Equipment**

  Properly installed to achieve operational efficiency; good quality equipment to be used. The equipment used should facilitate through cleaning; prevent physical and chemical change through contact and minimize contamination. The written instructions for utilization of equipment be provided and accuracy, precision should be maintained.

- **Raw Materials**

  Properly identified; analyszed; containers of raw materials inspected for any damage; stored at optimum temperature; labeled properly; systematically sampled by quality control personnel; tested for compliance of required standards; released from quarantine by quality control personnel through written instructions; and rejected materials destroyed or returned back to the supplier.

- **Personnel**

  Manufacture under direct supervision of competent technical staff; separate Head for Q.C. laboratory; qualified and experienced personnel for Quality Assurance and Quality Control Operations; written duties assigned; adequate number of personnel; good laboratory practices and; proper training of technical staff members.

- **Master Formula Records (MFR)**

  Licensee should maintain records relating to all manufacturing procedures for each product and batch size to be manufactured. It also includes patent or proprietary status; name of formulation alongwith generic name if any; name, quantity, and reference number of starting materials; strength; dosage form; description; identification; composition; statement of processing location; step-wise processing instructions; in-process control; requirements for storage conditions; packaging details, etc.

- **Batch Packaging Records**

  It is based on relevant parts of packaging instructions. Transcription errors to be avoided; packaging equipment should be clean; planned packaging operations and proper maintenance of packaging records.

- **Batch Processing Records (BPR)**

  BPR for each product; clean equipment; name of product; number and batch being manufactured; dates and time of commencement of operation; significant intermediate stages; initials of operator of different steps of production; batch number; analytical control number; in-process control records; amount of product obtained; note on any deviation from master formula; and addition of any recovered or reprocessed material.

- **Standard Operating Procedures (SOPs) and Records**

  SOP and records for receipts of each delivery of raw, primary and printed packing material; sampling; instrument and equipment; internal labeling; quarantine and storage; batch numbering; testing, records of analysis; equipment assembly and calibration; maintenance; cleaning and sanitation; personnel; pest control; and complaints, recalls made and returns received.

- **Manufacturing Operations and Controls**

  Competant technical staff supervision for weighing, measuring and other operations; nonsterile products free from *Escherichia coli* and *Salmonella* microbes; conspicuously labeled with name, batch number, and other details; cross-contamination avoided; and all process controls checked under master formula.

- **Reprocessing and Recovery**

  The reason for reprocessing should be specified, corrective measures for recovery should be spelt out only if permitted in Master Formula.

- **Product Containers**

  Compliance with pharmacopoeial requirements; cleaning procedures and sterilization procedure should be properly followed. There should be written schedule for programs for cleaning of containers. When bottles are not dried after washing, de-ionized water should be used for rinsing.

- **Labels and Other Printed Materials**

  Stored properly and separately; used as and when required and should not be inter-mixed.

- **Distribution Records**

  Records properly maintained; records of complaints, adverse reactions and other reactions from consumers should also be maintained.

- **Quality Control System**

  Detailed instructions for quality control of raw materials and finished product; quality control for packaging and labeling; adequacy of storage, quality control procedure revised as and when possible and qualitative examination of returned products.

PART II

(Salient Features) Plant and Equipment

Part II of Schedule M gives the details of the plant and equipment required for manufacture, quality control and quality assurance of different dosage forms. The specifications of equipments are also indicated. The details of requirements are categorized into eleven groups.

1. External Preparations

It covers ointments, emulsions, lotions, solutions, pastes, creams, dusting powders and other identical preparations.

(i) **Minimum area :** 30 square meters for basic installation and 10 square meters for ancillary area.

(ii) **Requirements :** mixing and storage tanks, jacketed kettles of different types, electric mixer, planetary mixer, colloid mill, triple roller mill, liquid and tube filling equipments, etc.

2. Oral Liquid Preparations

It covers syrups, elixirs, emulsions and suspensions.

(i) **Minimum area :** 30 square meters for basic installation and 10 square meters for ancillary area;

(ii) **Requirements :** SS mixing and storage tanks, jacketed kettles of different types, electric stirrer, electric colloidal mill, emulsifier, filtration equipment, bottle filling machine, cap sealing machine, de-ionizer or water distillation unit, clarity testing unit, etc.

3. Tablets

For effective production, tablet production department is divided into four sections:

(i) Mixing, granulation and drying section

(ii) Tablet compression section

(iii) Packaging section (strip/blister)

(iv) Coating section.

(a) **Minimum area :** A minimum of 60 square meters for basic installation and 20 square meters for ancillary area for un-coated tablets. For coated tablet, **additional** area of 30 square meters for coating section and 10 square meters for ancillary area.

(b) **Requirements :** Disintegrator, sifter, powder mixer, mass mixer, planetary mixer, rapid mixer granulator, granulator, hot air oven, weighing machines, compression machine (single, multi-punch, rotary), punches and dyer storage cabinets, table de-duster, table inspection unit/belt, dissolution test apparatus, single pan balance, hardness tester, friability and disintegration test apparatus, strip/blister packaging machine, leak test apparatus, tablet counter, jacketed kettles of different types, SS coating pan, polishing pan, weighing balance, exhaust system and vacuum dust collector, air-conditioning system (wherever applicable), etc.

4. Powders

(i) **Area :** Minimum 30 square meters; additional room for actual blending.

(ii) **Requirements :** Disintegrator, electric mixer, sifter, SS vessels and scoops of suitable sizes, filling equipment, weighing balance, etc.

5. Capsules

(i) **Area :** A separate enclosed area, suitably air-conditioned and dehumidified. A minimum area of 25 square meters for basic installation and 10 square meters for ancillary area each for penicillin and non-penicillin section.

(ii) **Requirements (for hard gelatin capsules):** Electrical mixing and blending equipment, capsule filling units (semi-automatic and automatic), capsules counters, weighing balance, disintegration test apparatus, capsule polishing equipment, etc.

6. Surgical Dressings

(i) **Area :** Minimum 30 square meters for basic installation; for medicated dressing additional room required.

(ii) **Requirements :** Rolling, staining, cutting, folding and pressing machines; mixing tanks, hot air oven, steam sterilizer, work tables, etc.

7. Ophthalmic Preparations

It includes eye-ointment, eye lotions and other preparations for external use. Separate enclosed areas with air-lock arrangements required.

(i) **Area :** Minimum 25 square meters for basic installation and 10 square meters for ancillary area.

(ii) **Requirements :** Hot air ovens, jacketed kettles of different types, colloid mill, ointment mill, SS-mixing and storage tanks; tube washing, drying, cleaning and filling machines; automatic vial washing machine, vial drying machines, sintered glass funnels, autoclave, liquid filling equipment, laminar flow units, air conditioning and dehumidification arrangement, rubber bung washing machine, etc.

8. Pessaries and Suppositories

(i) **Area :** Minimum 25 square meters for basic installation.

(ii) **Requirements :** Mixing, pouring and moulding equipments; weighing devices. For pessaries manufactured by granulation and compression, requirements shall be as given under "tablet".

9. Inhalers and Vitrallae

(i) **Area :** Minimum 25 square meters for basic installation.

(ii) **Requirements :** Mixing, graduated delivery and sealing equipments.

10. Repacking of Drugs and Pharmaceuticals

(i) **Area :** Minimum 30 square meters for basic installation. Exhaust system be provided in case of operations involving floating particles.

(ii) **Requirements :** Weighing, measuring and filling equipments; powder disintegrator, electrically operated powder sifter, electric sealing machine, SS scoops and vessels, etc.

11. Parenteral Preparations

The whole operation of manufacture (small volume injectables and large volume parenterals) in glass and plastic preparations is divided in separate areas/rooms.

(i) **Parenteral Preparations in glass containers**

It includes areas for water management, containers, closures preparation, solution preparation, filling, capping, sealing, sterilization, quarantine, visual inspection and packaging.

(a) **Area :** Minimum 150 square meters for basic installation and 100 square meters for ancillary area for small volume injectables.

(b) **Requirements :** Distillation unit, de-ionized water unit, thermostatically controlled water storage tank, transfer pumps, SS service lines for carrying water, automatic rotary ampoule/vial/bottle washing machine, automatic closures, washing machine, dryer, double ended sterilizer; storage equipment for ampoules, vials, bottles and closures, SS benches/stools, dust proof storage cabinets, mixing SS tanks, portable stirrer, filtration equipment, transfer pumps, automatic ampoule/vial/bottle filling, capping, sealing machines under laminar air flow work station; gas lines for nitrogen, oxygen and carbon dioxide; steam sterilizer, hot air sterilizer, storage cabinets, visual inspection units, batch coding, machine labeling unit, pressure leak test apparatus, etc.

For large volume parenterals the minimum area required is 150 square meters each for basic installation and ancillary area.

(ii) Parenteral Preparations in Plastic Containers by Form – Fill – Seal/Blow, Fill – Seal technology

The operational activities are in separate areas for water management, solution preparation, container-moulding-cum-filling, sealing, sterilization, quarantine, visual inspection and packaging.

(a) Area : Minimum 250 square meters for basic installation and 150 square meters for ancillary area. Areas for formulations meant for external and internal uses shall be separately provided. A minimum of 100 squares meters be provided for packaging materials for large volume parenterals.

(b) Requirements : De-ionized water treatment unit, distillation unit (multi-column with heat exchangers), thermostatically controlled water storage tank, transfer pumps, SS service lines for carrying water, storage tanks, solution preparation tanks, transfer pumps, cartridge and membrane filters, steril form-fill-seal machine, plastic granules feeding device, super-heated steam sterilizer, adequate number of platforms, racks for storage, visual inspection unit, pressure leak test apparatus, batch coding machine, labelling unit, etc.

Manufacture of Drugs

Definition : Manufacture in relation to any drug includes any process or part of a process for making, altering, finishing, packing, labeling, breaking up or otherwise treating or adopting any drug with a view to its sale and distribution, but does not include compounding or dispensing of any drug or packing of any drug in ordinary course of retail business.

Manufacture of drugs is a blend of art and science, to be achieved strictly in accordance with the provisions of Good Manufacturing Practices (GMP). A person who is interested in starting manufacturing of drugs is required to fulfill several conditions laid down in DCA and Rules. The conditions to be fulfilled before licence is granted are collectively called as **"Conditions Precedent"** and conditions that are required to be fulfilled after the licence is obtained for manufacturing are called **"Conditions Subsequent".** The Licensing Authority is both in States and at Central Government. The Central Government is empowered to prohibit manufacturing and sale of any drug formulation in public interest.

Licences are required for the manufacturing of following categories of drugs.

1. Manufacturing of drugs belonging to Schedules C and C (1)
2. Manufacturing of drugs belonging to Schedule X
3. Manufacturing of drugs belonging to Schedules C, C (1) and X
4. Manufacturing of drugs other than Schedule C, C_1 and X
5. Manufacturing of drugs for examination, test or analysis
6. Loan licences
7. Licence for repacking.

The following categories of drugs and cosmetics are **prohibited to be manufactured** or sold in our country.

1. Any drug or cosmetic which is substandard, misbranded, adulterated or spurious.
2. Any patent or proprietory medicine without clear indication of ingredients.
3. Any drug claiming for accurate cure or prevention of diseases listed in Schedule J.
4. Any manufacturing of formulation containing drug or cosmetic which has been imported into our country in contravention of the provisions of the Act and Rules.

5. Manufacturing for sale of any drug or cosmetic containing any harmful ingredient.
6. Manufacturing for sale of any drug or cosmetic in contravention of the provisions of the Act and Rules, provided that manufacture of small quantities of any drug for the purpose of examination, test or analysis is permitted, subject to prescribed conditions.

Separate applications for separate licences for more than one premises of manufacture are required to be made.

[A] Manufacturing of Drugs other than Schedules C and C (1)

Application for grant of licence or renewal is made in **Form 24.** The licence is issued by Licensing Authority in **Form 25.**

I Conditions for grant of Licence are as follows:

1. **Competent Staff:** (i) A graduate in Pharmacy/Pharmaceutical Chemistry with a minimum of eighteen months of experience after graduation. The duration may be reduced by six months, if the applicant has undergone training during graduation. **or** (ii) A graduate in science with chemistry as a principal subject and with three years of manufacturing experience after graduation **or** (iii) A graduate in chemical engineering or chemical technology or medicine with three years of manufacturing experience after graduation **or** (iv) holding equivalent foreign qualification.

   For disinfectant fluids, insecticides, liquid paraffin, non-chemical contraceptives, surgical dressings, medicinal gases, plaster of paris, only adequate experience in manufacturing is required and no specific qualification is mentioned.
2. **Factory premises :** As per Schedule 'M' with regard to premises, space, plant and equipment.
3. **Separate facilities for analysis** of raw materials and finished formulations.

   There should be separate head of department for analysis and manufacturing sections. Head of the testing and analytical department should be a graduate in medicine/pharmacy/

pharmaceutical chemistry/science with adequate experience in analysis.

4. There should be adequate arrangement of storage of raw materials and finished products.

5. While applying for licence to manufacture patent or proprietary medicines, it is required to submit evidence justifying therapeutic claims of the product, its stability and safety.

II Conditions to be fulfilled after getting a licence:

1. The manufacturer should always maintain adequate staff, sufficient premises and equipment.
2. The manufacturing records, records for raw material and analysis and other operational records should be maintained as per Schedule 'U'.
3. The licensee should own an analytical laboratory or get tested the samples analyzed in an approved analytical laboratory.

   The records for analysis are required to be maintained for a period of five years from date of manufacturing. The records are both for manufacturing and for finished product.
4. The manufacturer should allow the Inspector to inspect the premises, manufacturing process, analytical procedures, and withdraw the samples. The samples may be provided on demand and entire protocol of manufacturing should be made available when asked for. The manufacturer should also withdraw the batch manufactured by him if directed to do so by the Controlling Authority.
5. The manufacturer should comply with all the requirements of the Act and Rules thereunder.
6. The manufacturer should maintain the inspection Book.
7. Samples with expiry date should be maintained for six months after expiry date. For other categories, the samples should be maintained for three years from the date of manufacturing. Twice the lot of reference samples should be maintained. The quantity maintained should be sufficiently available for analysis.

8. Any change in the staff structure, especially technical staff, should be reported to Licencing Authority.
9. Any major structural change in the premises should be done with the permission of Licencing Authority,
10. The manufacturer should forward all the sales records to the Controlling Authority.

[B] Manufacture of Drugs for Testing, Analysis or Examination

If the manufacturer does not hold separate licence for test, analysis or examination, the licence is obtained in **Form 29.** The provisions relating prohibition of manufacturing of certain drugs do not apply for such manufacturing meant for test or analysis. The validity of the licence is for one year. The manufactured drugs should be kept in containers bearing appropriate label indicating the purpose of test or analysis. When the material is supplied to other manufacturer, the label stating the name and address of manufacturer, scientific name of the drug, licence number, date of manufacture, etc., should be provided.

The manufacturer should allow the inspector to inspect the premises, manufacturing, and analytical records and withdraw the samples if required for analysis. The manufacturer should comply with the provisions of the Act and Rules. The manufacturer should maintain an inspection book and the same be shown to the Inspector.

[C] Manufacturing of New Drugs

In addition to provisions for manufacture of drugs, there should be documentary evidence for quality, purity, therapeutic trials of new drugs and evidence for approval under schedule 'Y' (Clinical trials).

[D] Loan Licences

For drugs other than Schedules C, C (1) and X, loan licences can be given. A qualified person can make use of approved facilities of manufacturing provided by any other person and obtain a loan licence for manufacture of drugs other than Schedules C, C (1) and X. The licensee in such cases should convice the Licensing Authority about the availability of the infrastructure on loan from approved manufacturer, and also be convinced about the need to grant such a loan licence.

All the conditions of GMP in Schedule 'M' are required to be fulfilled. Manufacturing Records should be maintained for five years. In case of drugs with date of expiry, the records should be maintained for two years. Application for grant or renewal of loan licence is made in **Form 24-A.** The licence is issued by Licensing Authority in **Form 25-A,** which is valid for one year.

[E] Repacking Licence

It is issued for drugs other than Schedules C, C_1 and X, subject to fulfillment of conditions. The application is made for grant or renewal of licence in **Form 24-B.** The licence is issued by Licensing Authority after inspection in **Form 25-B.**

There should be adequate arrangement for testing of samples. The licence should always be displaced at premises of repacking. The factory primises for repacking should comply with provisions of Schedule M. Hygienic conditions of working should always be maintained. Adequate staff should be appointed and any change in staff structure should be immediately informed to Controlling Authority.

The Competent person eligible to get Repacking Licence is —

1. Diploma in pharmacy or Registered pharmacist
2. Intermediate with chemistry as a subject
3. Matriculation with four years of experience in manufacture, dispensing or repacking of drugs.

The licence is valid till 31st December every year and required to be renewed. There should be separate application for separate licence.

The container or package of repacked drug should bear on its label the words – "Rpg.Lic.No". The Repacking Licence is given to the competent persons for the bulk drugs or drug formulations procured in large quantities from manufacturer directly and required to be sold in small quantities.

Drug Related Activities

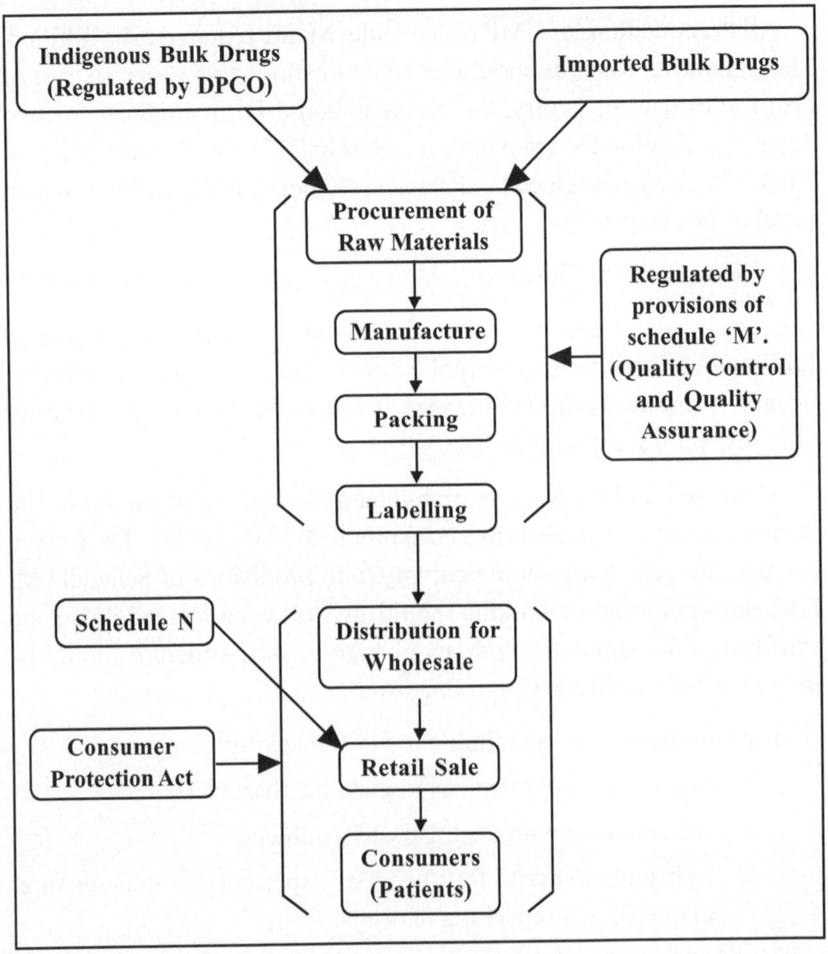

[F] Manufacture of Drugs under Schedules C and C (1) and Drugs specified in Schedules C, C (1) and X

Application for grant or renewal of a licence is made for Schedules C and C (1) drugs in **Form-27** and for Schedules C, C (1) and X in **Form 27-B.** The licence is issued in **Form-28** for manufacture of Schedules C and C (1) drugs and in **Form 28-B** for drugs under Schedules C, C (1) and X.

1. The conditions stipulated in Schedule 'M' should be met with. The requirements of space, plant, equipment, etc., should be as per Schedule M.
2. The testing strength or quality strength of the manufacturing unit should be assessed. There should be a qualified independent head for analytical wing including for analysis of Schedules C and C (1) drugs.
3. The competent technical persons should be available with one of the following qualifications.

   (i) A graduate in pharmacy/pharmaceutical chemistry with eighteen months of experience after graduation in manufacturing drugs to which licence applies. Six months of mandatory training after graduation can form part of this total training of eighteen months.

   or

   A graduate in science with chemistry or microbiology as major subject with a minimum of three years of experience after graduation in the manufacture of drugs to which this licence applies.

   or

   An equivalent foreign qualification or an expert in manufacturing before 29th June, 1957.

   For manufacture of veterinary Scheduled drug formulations, a graduate in medicine/veterinary science/general science/pharmacy with minimum three years experience in manufacture of biological products.
4. There should be facilities available for storage of drugs. Sufficient precaution should be taken for preserving the properties of manufactured drugs.
5. Records of manufacturing should be maintained as per Schedule 'U'. Records of test, analysis and examination and batch-wise records are required to be maintained. All records should be kept for a period of two years from the date of expiry of drug. In other cases, with no expiry date, the records should be kept for five years from the date of manufacturing.

6. The manufacturer should allow the inspector or the representative of Controlling Authority and Licensing authority to visit manufacturing premises with or without notice.
7. The manufacturer should report to the Licensing Authority any changes in staff or any material changes.
8. The licensee should furnish the sample of drugs to the inspector when asked for purpose of analysis. He should not sell the drug if asked to do so by Controling Authority and withdraw the drug already sold from the market on direction from the Authority.
9. The licensee should maintain the inspection book to enable the inspector to record his observations.

Ayurvedic, Siddha and Unani Drugs

Chapter IV A of the Act applies to Ayurvedic, Siddha and Unani Drugs. The related Rules governing activities are from Rules 151 to 167.

Ayurvedic, Siddha and Unani Drugs Technical Advisory Board (DTAB)

On lines with DTAB for Allopathic System of medicines, the Central Government constitutes the Advisory Committee called as **Ayurvedic, Siddha and Unani Drugs Technical Advisory Board.** The Board is required to advise Central and State Governments on all technical matters pertaining to DCA in relation to Ayurvedic, Siddha and Unani Tibbi medicines. Major differences as compared to that of DTAB (Allopathy) are as follows:

1. It consists of only Ex-officio and nominated members, whereas, in DTAB (allopathy) there are Ex-officio, nominated and elected members.
2. There are 20 members as compared to eighteen in DTAB (allopathy). This is because of three systems involved, namely Ayurveda, Siddha and Unani.
3. Chariman of this DTAB need not be always Director-General of medical and health services, Government of India.
4. The term of this DTAB is for three years.

Composition

The Central Government constitutes the Board with eighteen members representing different disciplines of Indigenous System of medicine (Table 8.1). The Central Government nominates one of these members as the Chairman of DTAB. The term is for three years. The Headquarters is at Nirman Bhavan, Ministry of Health and Family Welfare, Government of India. The Board meets usually twice a year to discuss on technical matters pertaining to Ayurveda, Siddha and Unani Medicines in relation to DCA. The Central Government appoints Secretary, Assistant-Secretary and other required staff.

Ex-officio Members (4)	**Nominated Members (16) All nominated by Central Government**
1. Director General of Medical and Health Services, Government of India	(i) One Government Analyst (ii) One Pharmacognosist (iii) One Phytochemist
2. Drugs Controller General of India	(iv) One Teacher / in Dravyaguna and Bhaishaja Kalpana
3. Director, Central Drugs Laboratory, Kolkata	(v) One Teacher in ILM – UL – ADVIA and TAKLIS WADAWASAZI
4. Principal Officer incharge for Indigenous System of Medicine in Ministry of Health and Family Welfare, Government of India	(vi) One Teacher in Gunapadam (vii) Three persons from industry representing Ayurvedic, Siddha and Unani industrial units (one each) (viii) Three persons one each from among the practitioners of Ayurvedic, Unani and Siddha Systems of Medicine (ix) Four persons to be nominated, two from Ayurvedic Pharmacopoeia Committee, one from Siddha Pharmacopoeia Committee and one from Unani Pharmacopoeia Committee.

Ayurvedic, Siddha and Unani Drugs Consultative Committee

On lines with Drugs Consultative Committee for allopathic drugs, there is a separate Drugs Consultative Committee for Ayurvedic, Siddha and Unani Drugs. It is a Statutory Committee charged with the responsibility of advising Central Government and State Government, as well as, Drugs Technical Advisory Board for Ayurveda, Siddha and Unani Drugs on all matters pertaining to uniform implementation of provisions of the Act and Rules in relation to Ayurvedic, Siddha and Unani drug formulations. The Committee comprises nominated members, two from Central Government and one each from State Government. Drugs Consultative Committee meets as and when required. The Director General of Medical and Health Services is usually the Chairman of Drugs Consultative Committee.

Prohibited from Manufacture

- No person shall manufacture (i) misbranded, adulterated or spurious Ayurvedic, Siddha or Unani drug and (ii) patent or proprietary medicine unless displayed in prescribed manner with list of ingredients and sell, stock or exhibit any such drug manufactured in contravention with provisions of the Act and Rules. However, Vaidyas and Hakims may manufacture drugs for their own patients.
- The Central Government reserves power to prohibit, manufacture etc., of Ayurvedic, Siddha or Unani drugs in public interest.

Manufacture of Ayurvedic, Siddha and Unani Drugs

The State Government appoints Licensing Authority for these drugs. Separate application for separate licence for each premises is required to be made.

Form 24-D - Application for grant or renewal of licence

Form 24-E - Application for loan licence

Form 25-D - Issue of licence to manufacture

Form 25-E - Loan licence for manufacture

Form 35 - Inspection book to be maintained

Form 48 - Approval for carrying out tests on behalf of licensee for manufacture.

The Good Manufacturing Practices (GMP) for Ayurvedic Drugs have been introduced in 2000. GMP covers requirements of factory premises, equipment for manufacture and analysis, storage, etc.

The licence for manufacture is issued within three months from date of receipt of application. The licence is granted after consulting an expert identified by State Government. The licence for manufacture and loan licence are valid up to 31st December of the year following the year in which it is granted or renewed.

The licensee or loan licensee shall maintain proper records of manufacture and analysis, allow Inspector to inspect premises and maintain an inspection book in **Form 35**.

The Licensing Authority, after giving an opportunity to show cause, after fifteen days, may cancel a licence or suspend it for a specific period, if the licensee fails to comply with conditions of licence. The aggrieved person may appeal to the State Government within three months which shall take final decision.

Labeling and Packing

The label should contain:

(i) Mention of Ayurvedic / Siddha /Unani medicine
(ii) List of ingredients
(iii) Weight, measure or number, as the case may be
(iv) Name and address of manufacturer
(v) Licence number being preceded by the words 'Manufacturing Licence Number' or 'MFg. Lic. No.' or 'M.L'
(vi) For internal use of schedule E (1) drugs on container-, 'Caution-to be taken under medical supervision' both in English and Hindi
(vii) A distinctive batch number
(viii) Date of manufacture

Exemption

Label and packages or containers of Ayurvedic, Siddha and Unani drugs for export may be adapted to meet specific requirements of the law of the country importing such drugs.

State Drug Licensing Authority shall have Ayurvedic / Siddha / Unani qualifications as per schedule II of CCIM Act 1970/B. Pharm. (Ayurveda) of a recognized university with minimum five years experience in manufacture or testing of such drugs.

The Government Analyst is one appointed under section 33-F of the Act and shall be a person possessing qualifications prescribed in Rule 44 or a degree in Ayurveda, Siddha and Unani system with minimum of three years of post-graduate experience in analysis of drugs in a laboratory under control of Government Analyst or a Chemical Examiner or Head of approved institution.

The Qualification of Inspector is (i) a degree or diploma in Ayurvedic or Siddha or Unani system or a degree in Ayurvedic Pharmacy or (ii) a qualification laid down under Rule 49 and shall have undergone practical training in manufacture of Ayurvedic, Siddha or Unani drugs.

Standards

The drugs included in Ayurvedic Pharmacopoeia have to comply with standards for identity, purity and strength given in it.

Asavas and **Aristas** have to comply with upper limit of alcohol (self-generated alcohol should not exceed 12 % v/v).

Special Provisions Relating to Biological and other Special Products

These products cannot be sold unless they have been sealed in previously sterilized containers made up of glass or other suitable materials approved by the Licensing Authority. This is not necessary if the substance is to be terminally sterilized.

In multiple dose containers, sufficient proportion of antiseptic/ preservative is to be added. This is not applicable to penicillin suspensions. Test for sterility for surgical ligatures and sutures should be carried out as per Schedule F.

Special provisions of biological products can be covered under the following.

1. Test for sterility
2. Test for freedom from abnormal toxicity
3. Test for pyrogens
4. Approval of testing institutions for biological products.

Test for sterility, abnormal toxicity and pyrogens are carried out as per procedures described in Indian Pharmacopoeia.

Following biological and other products shall be tested for freedom from living aerobic or anaerobic microorganisms.

(a) Sera and solution of sera proteins

(b) Bacterial vaccines and anti-rabies vaccine

(c) Preparation of posterior pituitary lobe

(d) Toxins, antigens, and their mixtures and immunization or diagnostic products

(e) Preparation of insulin

(f) Any other parenteral preparation

(g) Any preparation from culture of pathogenic organism to be administered orally which must be sterilized.

Procedure for sterility test (some requirements)

1. **Preparation:** Not less than 0.1% volume of batch for a batch which is less than 10 litres; 10ml if volume is more than 10 litres.
2. **Containers:** Not less than 1% of total number of containers if the batch is of more than 1000 containers; and a minimum of 10 containers, if the total number is less than 100.
3. **Medium for aerobic microorganism:** It consists of either meat extract with addition of 1% peptone or equivalent medium.
4. **Medium for anaerobic microorganism:** It consists of nutrient broth similar to use in testing aerobic microorganism with addition of heat coagulated musle, sufficient to occupy a depth of not less than 1 cm at the bottom of the tube.

For manufacturing of Schedules C, C (1) and X, the particulars of production, quantity of Schedule X drugs used for manufacturing and quantity of manufactured drugs are required to be maintained by the licensee.

Packing and Labeling of Drug Formulations

Different steps in the manufacturing of drugs are suitably monitored by way of provisions made under DCA and Rules. The stages involved from procurement of raw material to the sale of drug formulations at the retail counters are mandatorily controlled in accordance with the provisions of the Act. In addition to mandatory requirement, it is also the moral, ethical and social responsibility of the manufacturer to ensure that the consumer receives good quality of his/her products. Packing is a blend of art and science with regulatory flavor. It is not sufficient to provide artistic packaging of the formulations but, it should be sufficient enough to ensure the stability of product during transportation and storage, assuring high quality of the product.

The text of labeling on packing material varies with the type of product formulated. There are specific requirements of labeling for drugs of Schedules G , H and X; external applications, patent and proprietary medicines, ophthalmic ointments, contraceptives, disinfectants and several other drug formulations.

Labeling should be attractive and readable. It should be in printed form on the outerside of the packing material, as well as on the packing of drug formulation. Even single unit of formulation (ampoule or tablet) should have appropriate label on it. In case of single dose of tablet, it could be short name of the product embossed on it.

The following particulars should appear in the label of the drug formulation.

1. Name (Patent or Proprietary and Generic name).
2. Name and address of manufacturer.
3. Batch or lot number.
4. Date of manufacturing.
5. Expiry date, if any.
6. Information for storage, if any.
7. Precautionary information - i.e., care in handling the product, use, etc.
8. General information including – Physicians sample – not to be sold, in case of physicians.

In case of medicines made up ready for treatment, name and address of licensee by whom it is supplied should appear on the label.

In case of preparations included in B.P, B.P.C, I.P the abbreviations should be mentioned.

In a preparation containing more than 3% alcohol, the percentage of alcohol should be mentioned on the label.

Particulars of Label

Category	Important particulars on label
1. **Schedule H drug (Prescription drug)**	(i) R_x on left top corner of label. (ii) "Schedule H drug - Warning : To be sold by retail on prescription of RMP only". This text should appear on label.
2. **Schedule G drug**	**Caution** – "It is dangerous to take this prescription except under medical supervision". This should be surrounded by a line within which no other word should be printed.
3. **Schedule X drugs made**	(i) NR_x in red ink on left top **up for internal use** corner of label. (ii) Schedule X drug – warning - to be sold by retail on prescription of RMP only.
4. **Schedule X drug in bulk form**	Should have XR_x in red ink on the label.
5. **Schedule C drug**	(i) Proper name of the drug in addition to patent or proprietary name (ii) Licence number (iii) Batch/Lot number (iv) Statement of potency in units (v) Manufacturer's name and address (vi) Date of manufacturing (vii) Statement for test for maximum toxicity (viii) Date of expiry (ix) Nature and percent of antiseptic or preservative added (x) Precautions for preservation
6. **Patent/Proprietary medicine containing vitamins for prophylactic or therapeutic use.**	Should have on its label words: "For prophylactic use" or "For therapeutic use".

Contd....

Category	Important particulars on label
7. Non sterile ligatures and sutures	"Non-sterile surgical ligature or suture - Not to be used for operation upon human body unless effectively sterilized" and this part of label should be in red ink.
8. Pharmacopoeial and other drugs	(i) It should have on label words printed as I.P., U.S.P., and B. P. (ii) Net amount of drug in metric system. (iii) Amoun1t of active ingredient and all other particulars.
9. Ophthalmic Ointment	(i) It should have warning on label: If irritation persists or increases discontinue the use and consult physician. (ii) Special instructions for storage, if applicable.
10. Medicines containing methylated spirit	It should have words "For external use only".
11. Disinfectants	Name of the disinfectant, grade, type, potency, quantity in container, ndications for mode of use and adress of manufacturer.
12. Oral contraceptives	Date of manufacturing and instructions for usage.
13. Colored medicament or formulation	Common name of the colour used and its percentage (only permitted colors are to be used).
14. Medicines for veterinary purpose	Symbol of animal-head depicted on label, "Not for human use–for animal treatment only".

The label should be printed or written in indelible ink and should clearly appear on label of inner most container and every other covering of the container. The details of labeling include :

1. Name of the drug (Trade name/Generic name as applicable) should be printed clearly.
2. Net contents (weight, measure, volume or number of units of activity as applicable). Net content should be in metric system.

3. Contents of active ingredients:
   (i) For solid oral dosage forms (tablet, capsule), content in each unit of formulation.
   (ii) Solid form for injectables, in terms of weight mg/gm of powder. In case antibiotic, it is in terms of units of activity.
   (iii) Liquid orals: contents of ingredients in single dose of 5 ml or multiple. If dose is less, the contents of ingredients in 1 ml of preparation.
   (iv) Liquid parentral preparation: contents of ingredients in 1 ml or per dose in case of single dose preparation.
4. Name and address of the manufacturer on small container. Name and place of manufacture is sufficient on each ampoule.
5. Manufacturing licence number abbreviated as Mfg. Lic. No.
6. Distinctive Batch No. should be written as Batch No. or B.No. or Batch, Lot No.
7. Date of manufacturing.
8. Expiry particulars, if any.
9. Precautions related to handling, use or distribution.
10. Information on storage.
11. Any other general information or specific information pertaining to formulation. If the sample is for physician, the words "Physicians sample – not for sale" should be printed.

For Schedules F and F (I) and X drugs, only Code No. as approved by the Licensing Authority is required to be printed.

Specific Requirements

Biological and other specific products of Schedules C and C (1) advertized and sold as proprietary medicine, proper name of the substance should appear on label of every ampoule or container. Proper name, Manufacturing Licence Number, Batch Number, statement of potency in units, expiry date, storage requirement, etc., should be mentioned.

For ophthalmic solutions and suspensions, the additional requirement on lable is: "Use within one month of opening and not for injection."

The standard reference books are I.P, US.P, B.P, B.P.C, National Formulary of U.S., State Formulary of U.S.A, International Pharmacopoeia and Pharmacopoeia of Soviet Socialist Republic.

Mechanical Contraceptives

The label should contain –

1. Particulars specified in Schedule R,
2. Date of manufacturing,
3. Date up to which it retains its property,
4. Storage, in addition to other information.

Oral Contraceptives

In addition to general information, label should depict – date of manufacture and usage.

Drugs for export should be labeled as per rules prevailing in country of export or importing country. In Schedule E (1) drugs, label in red printing or printed against red background.

Labeling and Packaging of Ayurvedic and Unani Drugs

The label should be in indelible ink and clear, and it should reflect following particulars.

1. True list of all ingredients with quantity in metric system.
2. Reference to the method of preparation, name of authoritative book in First Schedule. If the list of ingredients is very large, it should be printed and packed in the container.
3. Name of the drug as per literature in the First Schedule.
4. If poisonous or schedule E (1) drug is contained, the words to be printed on label are "Caution – to be taken under medical supervision".
5. Name and address of manufacturer.
6. Date of manufacturing.
7. MFg. Lic. No (Manufacturing Licence Number).
8. Batch No. or Lot No.
9. Words - **"Ayurvedic Medicine"** or **"Siddha Medicine"** or **"Unani Medicine"**, as the case may be.
10. The drugs for external application should have the words on label "For external use only".
11. "Physician's sample not to be sold", if the sample is for physician.

Provisions Pertaining to Homoeopathic Medicines

Homoeopathic medicines are manufactured on the basis of long clinical experience in homoeopathy or as recorded in Authoritative Homoeopathic Literature in India and abroad.

No parenteral homoeopathic preparations are permitted for manufacture. **Import** – The import of homoeopathic medicine is without licence. The licence is necessary only for new homoeopathic medicine. i.e., the medicines not specified in homoeopathic pharmacopoeia of India, UK, U.S.A, and Germany.

Manufacture-For manufacturing each set of homoeopathic medicine, separate licence is required. Application for grant or renewal of a manufacturing licence is made in **Form 24-C.** Licence is issued in **Form 25 C.** Schedule M and GMP should be followed for manufacture of homoeopathic drugs. There should be competent staff, facility for manufacture and analysis, at least one permanent employee with a minimum of five years of experience in manufacturing of homoeopathic medicines. Factory premises should be clean and hygienic. The testing facility of Mother Tincture and facilities for proper storage of Mother Tincture should be provided. Mother Tincture should be stored in neutral clean glass containers. The manufacturer should maintain the register with all relevant particulars and allow the Inspector to inspect the premises and record his observations. The crude drugs used in preparation of Mother Tincture should be identified for their authenticity. The total solids and alcohol contents of homoeopathic preparations should be determined, recorded and the records maintained for five years.

No color should be added to the preparation, except in combination with syrup base.

Sale: Application for licence to sell, stock or exhibit or distribute homoeopathic medicine is made in **Form 19-B.** The licence for retail sale is issued in **Form 20-C.** The wholesale licence for homoeopathic medicines is issued in **Form 20-D.** Separate licence for separate place of retail is required. Sale premises should be clean. Incharge of homoeopathic medicines should be a competent person with five years experience. Homeo-doctor cannot look after premises for sale. The records of sale should be maintained properly in accordance with provisions of Rules. The Inspector should be allowed to inspect premises of sale.

No records of retail sale are required to be maintained, if the container of less than 30 ml of medicine and Mother Tincture made up to 60 ml potency. The records of sale for other categories should be maintained with particulars such as percentage of alcohol in preparation, name and address of manufacturer, date of manufacturing, manufacturing licence number, batch number, etc.

Provisions Pertaining to Cosmetics

Import of Cosmetics

Import of cosmetics of following categories is prohibited into our country.

1. Misbranded, spurious, adulterated and substandard cosmetics.
2. Cosmetics containing any ingredient which may render it unsafe for use.
3. Cosmetics whose import is prohibited under the provisions of the Act and Rules.
4. Cosmetic containing coal-tar color, other than one mentioned in Schedule.
5. Cosmetic containing more than 2 ppm of arsenic, 20 ppm of lead or 100 ppm of any metal.
6. Cosmetic intended for use on eye brows or eye lashes which contains coal-tar dye color, or intermediate or a base.
7. Cosmetics containing hexachlorophene.
8. Cosmetics containing mercury compounds.

Manufacturing of Cosmetics

Application for grant or renewal of a licence to manufacture cosmetic is made in **Form 31** and the licence to manufacture is issued in **Form 32**. Schedule M and GMP, competent staff, hygienic conditions of working, adequate space, plant structure and good testing facilities should be in accordance with the provisions of the Act and Rules. The licence should be displayed in the manufacturing premises. Any change in staff should be informed to Licensing Authority.

The loan licence for manufacture of cosmetics is issued in **Form 32-A** for which application is made in **Form 31-A**. For loan licence, all the conditions of space, plant structure, testing facilities, maintenance of production and analytical records should be fulfilled. The Licences are valid till 31st, December every year and are required to be renewed.

Labeling of Cosmetics

It should be both on inner and outer label and should contain name of cosmetic, name of manufacturer and place of manufacture. The outer label should contain net content of cosmetic i.e., weight for solids, measure for liquids and weight for semi-solids. On inner label of cosmetic with hazardous ingredient, adequate direction should be printed in the form of warning. The statement of quantity of hazardous ingredients, Mfg. Lic. No., Batch No. and other particulars should be depicted on the label.

For hair dyes containing coal tars

On inner and outer labels, the words printed should be "Caution – the product contains ingredients which may cause skin irritation in certain cases, and a preliminary test according to accompanying direction should be carried out. This product should not be used for dyeing eye-lashes or eye-brows as such use may cause blindness". This statutory warning should be printed both in English and local language.

The cosmetic should conform to the standards laid down by Indian Standards Institution.

Sale of Cosmetics

The cosmetics which do not contravene the provisions of Rules can be sold without licence. It is prohibited to stock misbranded, spurious, substandard and adulterated cosmetics. The dealer and shopkeeper should reveal the particulars of the stock of cosmetics, if asked by the Licensing Authority or Controlling Authority.

Sale of Drugs

The ultimate objective of pharmaceutical manufacture is to sell quality drug formulations to the consumers or patients. Pharmaceutical marketing today is a glittering and challenging professional activity. The distribution channel in pharmaceutical marketing is –

(i) Manufacturer → Distributor → Wholesaler → Retailer → Consumer

(ii) Manufacturer → Wholesaler → Retailer → Consumer

Definitions

Pharmacy

It includes every store, shop or any other place where -

(i) drugs are dispensed i.e., measured, weighed, made up to and supplied, or

(ii) drugs are prepared, or

(iii) prescriptions are compounded or a place which by sign/symbol or indication gives the impression that operations mentioned in (i) or (ii) as above are carried on in the premises where the words "Pharmacy" or "Pharmacist" or "Dispensing Chemist", or "Pharmaceutical Chemist" are displayed.

Drug Store : A place for retail where the licensee does not engage a qualified person or registered pharmacist.

Chemist and Druggist : A retail shop where the licensee employs qualified person or registered pharmacist, but does not maintain pharmacy.

Pharmacy, Dispensing Chemist, Pharmaceutical Chemist

A place where the licensee employs qualified person or registered pharmacist and also maintains a pharmacy.

Licensing Authority should look into average number of licences granted in the area, occupation and trade of the applicant before issuing licence for a period of three years.

The licences are issued for wholesale and retail shops. The specified job requires specified licence. Licensee should maintain register with necessary particulars as required under the Rules. Licence in all cases is required to be displayed at a prominent place in the premises. The licence for wholesale or retail should comply with provisions of the Act and Rules.

Wholesale of Drug

The wholesale trade is permitted to be carried out by one of the following categories of persons under Rule 64, Sub-Rule (ii). The wholesaler who is qualified is called a competent person, who is either (a) registered pharmacist (b) matriculate with four years experience in handling or sale of drugs, and (c) any graduate with one year experience in sale of drugs.

In addition to the requirements already mentioned, specific requirements are as follows:

1. For drugs of Schedules C and C (1), the licence for wholesale is issued in **From 21-B** by the Licensing Authority. For sale of each category of drugs, permission should be obtained from Licensing Authority. The storage of drugs should be in accordance with Schedule N.
2. For Schedule X drugs, the licence for wholesale is issued in Form 20-G. The licensee should forward to Licensing Authority copies of invoices of sale made to retail dealers.
3. For drugs other than Schedules C, C (1) and X, the licence for wholesale is issued in **Form 20-B**. The drugs are purchased from duly licenced manufacturer or dealer and should be sold to the person or a retailer holding requisite qualification. However, supply to hospitals, medical, educational and research institutions can be made directly by the wholesaler. Direct supply to RMP is also permitted.

Wholesale for Other than Schedules C, and C (1) from Motor Vehicles

Licence for wholesale of this category of drugs is issued in **Form 20-BB.** All conditions are required to be met with as stipulated earlier. The licensee should inform in writing any change in ownership of vehicle within seven days to Licensing Authority.

Retail Sale

The retail trade is solely in hands of registered pharmacists or qualified persons.

The licensee should maintain registers with following particulars:

(A) Drugs Other than Schedule X on a prescription of RMP

1. Serial number of entry
2. Date of supply
3. Name of the prescriber
4. Name and address of the patient/owner of animal in case of veterinary drug
5. Name and quantity of drug

6. For Schedules H and X drugs - name of the manufacturer, its batch number and expiry date
7. Signature of the qualified person under whose supervision the drug is supplied.

If not compounded in premises of the retail shop and supplied in original container, the particulars under 1-6 above should be mentioned in cash or credit memo.

No record for prescription against Employees State Insurance Scheme and other related schemes.

(B) Drugs belonging to Schedules C and C (1)

The requirement given under 1, 2, 4, 5, 7. as given above under (A) are also for this category of drugs. In addition, name of the manufacturer, batch no., date of expiry are required to be mentioned.

All above entries are required to be made in cash or credit memo book.

(C) Drugs under Schedules H and X

Schedule H prescription drugs: These drugs should not be sold except on prescription of RMP. For sale of Schedule X drugs, the prescription should be accepted in duplicate. One copy of this prescription is required to be retained by the licensee for two years. The prescription accepted should be in clear writing and signed by the prescriber/doctor with date. The name of the patient or owner of the animal, if it is veterinary drug, should also be indicated by the doctor. The information in prescription should also cover the total amount of medicine to be supplied and the dose to be given. The instructions in writing by the superintendent of hospital or RMP should only be honoured provided they are in accordance with the Rules. The records and the copy of the prescription should be retained by the licensee for a period of at least two years.

Schedules H and X drugs should be kept under lock and key in a separate cupboard. The pharmacist should not supply any substitute to these drugs. Schedule X drugs are recorded serially in the register. Separate page for each drug should be maintained in the register with following details.

(i) Date of purchase by the retailer, (ii) quantity received, (iii) name and address of supplier or wholesaler, (iv) name and quantity of drug supplied to the patient on prescription of RMP, (v) manufacturer's name, (vi) batch number, (vii) lot no., (viii) name and address of patient to whom it is supplied, (ix) reference No. of prescription against which it is prescribed, (x) name of doctor, (xi) bill number issued, (xii) date of receipt issued, and (xiii) signature of qualified person.

Storage of Veterinary Drugs

A separate cupboard or drawer should be reserved for veterinary drugs in the premises with no access to the customer. The drugs should be kept under lock and key.

Cancellation and Suspension of Licence

In case of violation of conditions of licence issued, the Licensing Authority, after giving sufficient opportunity to the licensee to explain his/her stand, may cancel or suspend wholly or partly, the licence for the wholesale or retail of drugs. However, if it is proved later that the violation of conditions of licence is not due to direct involvement of the licensee or not instigated by him/her or the licensee is not guilty of any serious act within twelve months and if it is proved that he/she was sincere and violation was not intentional, he/she may apply for reconsideration to the Licensing Authority of the State Government within three months from the date of suspension or cancellation of licence.

Mere storage of expiry samples in the premises is no offence, unless it is given or sold.

Offences and Penalties

(A) Manufacturing/Sale of drugs

1. Adulterated or spurious drugs with toxic substances:

   A minimum of five years imprisonment, which may extend up to life imprisonment and a fine of minimum ₹ 10,000/- for first conviction. For subsequent conviction, imprisonment up to ten years and may extend to life or fine of ₹ 20,000/- or both.

2. Adulterated drug formulation but not containing toxic substances or drug manufactured without license: Imprisonment from 1-3 years and fine of ₹ 5,000/. The Court may, however, reduce the punishment for first conviction. For subsequent conviction, imprisonment from 2-4 years and fine of ₹ 10,000/-. The Court may reduce the quantum of punishment.
3. Any other contravention of Drugs and Cosmetics Act and Rules: For first conviction, imprisonment from 1-2 years and fine. The Court may reduce the punishment. For subsequent conviction, imprisonment for 2-4 years and fine of ₹ 5,000/- or both.
4. For offence involving not disclosing the name of manufacturer or place, imprisonment up to three years or fine of ₹ 1000/- or both in first and subsequent convictions.
5. For an offence of not keeping records in order, imprisonment up to three years or fine of ₹1000/- or both.
6. Offence involving use of the report of Government analyst for advertising:- First conviction is punishable with fine up to ₹ 500/- and for subsequent conviction, imprisonment up to ten years or fine or both.

(B) Cosmetics

1. For spurious adulterated cosmetics, imprisonment up to 3 years and fine.
2. For not keeping records of cosmetics properly, imprisonment up to three years and fine up to ₹ 5,000/- for manufacturer and wholesaler.

Licences for Sale

Form 20 - Licence for retail sale of drugs other than Schedules C, C (1) and X.

Form 20-A - Restricted retail licence for drugs other than Schedules C, C (1) and X.

Form 20-B - Wholesale licence for drugs other than Schedules C, C (1) and X.

Form 20-F - Retail licence for Schedules X drugs.

Form 20-G	-	Wholesale licence for Schedule X drugs.
Form 20-BB	-	Wholesale licence for drugs other than Schedules C and C (1) from motor vehicle.
Form 21	-	Retail licence for Schedules C and C (1) drugs, excluding Schedule X.
Form 21-A	-	Restricted retail licence for Schedule C and C (1) drugs, excluding Schedule X.
Form 21-B	-	Wholesale licence for Schedule C and C (1) drugs.
Form 21-BB	-	Wholesale licence from motors vechicle for Schedules C and C (1) drugs.

CHAPTER 9

Good Clinical Practice (GCP)

Good Clinical Practice (GCP) is defined as a strategic approach for designing, conducting, monitoring, recording, analysing, auditing, and reporting of clinical trials that provides assurance that the data generated and results reported are credible and accurate, and that the rights, integrity and confidentiality of trial subjects are protected.

GCP protects the interest of the subjects in clinical studies. Adherence to the principles of GCP is very important. Otherwise, subjects involved in the clinical trials may be put at risk. It is also possible that data generated from clinical studies and submitted to the health authorities and scientific committee may be rejected, if found to be unreliable. The research credibility of the researcher and the research institution may be damaged, if ethical approach to clinical trials is not adopted. It is necessary that investigators with GCP training and certification should be chosen by the drug or healthcare companies to ensure that the studies conducted conform to regulatory requirements.

Historical Perspective

The shocking reports of the post Second World War (1939-45) trials of German medical practitioners accused of performing experiments on human participants without their knowledge and exposing them to the risk of death or permanent impairment of their faculties raised serious concerns about clinical studies. This prompted to come out with the first international statement on the ethics of medical research using human subjects namely, the **Nuremberg Code** in 1947.

Historical Milestones

1947 - **Nuremberg Code** was formulated for essentiality of voluntariness of the consent of the human Subjects.

1948 - U.N.General Assembly adopted Universal Declaration of Human Rights.

1964 - **Helsinki Declaration** The Council for International Organization of Medical Sciences (CIOMS) at Helsinki formulated guidelines for human use in medical research.

1966 - International Convention on Civil and Political Rights Stated that no one shall be subjected without his/her consent to medical or scientific treatment.

1980 - The Indian Council of Medical Research (ICMR) released Policy Statement on ethical considerations involved in research on human subjects.

1982 - WHO and CIOMS issued the proposed international guidelines for bio-medical research involving human subjects.

1990 - A set of European GCP guidelines followed.

1991 - CIOMS published international guidelines for ethical review in epidemiological studies.

1991 - GCP guidelines developed in Japan.

1993 - CIOMS issued international ethical guidelines for biomedical research involving human subjects.

1996 - ICH-GCP came into existence.

1997 - UNESCO brought out the Universal Declaration on Human Genome and Human Rights.

1997 - European Union and Japan implemented ICH-GCP guidelines.

1997 - US-FDA notified requirement of ICH-GCP for new drug application (NDA).

2000 - Report of National Bioethics Advisory Committee, USA published.

2002 - Guidelines issued by Nuffield Council of Bioethics, U.K. and CIOMS, Geneva.

2003 - International Declaration on Human Gene Data by UNESCO.

2005 - Universal Declaration on Bioethics and Human Rights by UNESCO.

The United States Food and Drug Administration (US-FDA) implemented guidelines for clinical investigators in mid 1970s which was followed by enforcement of rigorous investigational new drug (IND) procedures. These procedures were regularly updated over the years to the version that is in force today. Due to strict implementation of the new regulations in USA, the FDA could reject the data of substandard quality from other countries. These developments left other countries with little choice but to review their own procedures resulting in the birth of various national GCPs. In 1990, a set of European GCP Guidelines were followed and implemented. Japan, the 3rd largest pharmaceutical market, developed its own GCP Guidelines in 1991. However, the development of different GCPs did not solve the problem of acceptability of data as there were differences in GCP procedures in Europe, USA and Japan. As a result of this variation in GCP guidelines, the data generated in one country would not be automatically accepted by another country.

A series of meetings held amongst the officials of regulatory authorities and the representatives of various pharmaceutical companies in the USA, Europe and Japan, together with observers from Scandinavia, Australia, Canada and the WHO **(collectively known as ICH)** resulted in emergence of a set of GCP guidelines that were to be universally accepted. **In May 1996, the ICH GCPs came into existence and they are now the standards by which all clinical trials have to be conducted in order to achieve universal recognition.** In 1997, the European Union and Japan accepted the ICH GCP as a legal requirement. From May 1997, the US-FDA expects all clinical trials conducted outside the USA and used to support an application for marketing authorisation i.e. New Drug Application (NDA) in USA are conducted in accordance with the ICH GCP Guidelines.

ICH-GCP

The ICH Harmonized Tripartite Guidelines for GCP are broadly categorized as follows:

1. Rights and protection of subjects
2. Trial
3. Protocol
4. Roles, responsibilities, qualifications and experience
5. Procedures
6. Investigational Products

1. Rights and Protection of Subjects

The rights, safety and well being of the trial subjects are the most important considerations and should prevail over interests of science and society. Any trial taken must include freely-given consent of subject for taking part in a clinical study. To protect the rights of subjects, all clinical trials are required to be conducted in accordance with the ethical principles based on the Declaration of Helsinki. They should be consistent with the applicable regulatory requirements. The clinical trials undertaken should also weigh foreseeable risks and inconveniences against the anticipated benefits for the individual trial subject and society. A clinical trial should be initiated and continued only if, the anticipated benefits justify the risks.

The written informed consent must be obtained from each patient in accordance with regulatory requirements and the Declaration of Helsinki. Both the person taking the consent and the patient should personally sign (or thumb print) and date the consent form in the presence of a witness present during the whole consent procedure. The witness is also required to sign.

The volunteers for clinical trials can withdraw from studies at any time without giving a reason and such withdrawal does not prejudice any future medical treatment. It is necessary that the subjects understand the purpose of the study, how it affects them and what are the responsibilities of the sponsor and the investigators. The subject should be provided with this information both verbally and in writing. The information sheet should be provided for each subject. A 20-point check list for obtaining informed consent in accordance with ICH GCPs is made available. The subjects should also be informed of any compensation in the event of injury or death. The sponsor is required to provide indemnity insurance to cover the liability of the investigator and the sponsor in the event of any untoward event that may occur in subjects taking part in the study. The confidentiality of records related to the subjects must be maintained, respecting his/her privacy and confidentiality in accordance with the applicable regulatory requirements.

2. The Clinical Trial

The proposed clinical trial is required to be supported by adequate non-clinical and clinical information on the investigational product under study. The information should be scientifically acceptable and described in a clear detailed protocol.

3. The Protocol

The protocol refers to the formal record of how the trials must be conducted. It should be clearly understood and detailed. The protocol becomes operational only after the approval of the Institutional Research Review Committee and its Ethics Committee. No modification can be made without prior consultation with the sponsor, unless it poses immediate danger to subjects under investigation. All such agreed amendments to the protocol must be documented and approved by the Ethics committee (unless purely administrative in nature) prior to implementation. The composition of the Ethics committee must be in accordance with ICH-GCP requirements. The Ethics Committee should safeguard the rights, safety and well-being of all trial subjects.

4. Roles, Responsibilities, Qualifications and Experience

The medical care provided to and medical decisions made on behalf of subjects throughout the clinical trials shall always be the responsibility of a qualified physician or, when appropriate, of a qualified dentist. All investigators involved in the clinical trial must be qualified by education, training and experience to perform his/her respective tasks. The investigator should provide the sponsors with an up-to date, signed and dated CV as evidence of his/her qualifications, alongwith PG qualification and publications, if any. The investigator should be disciplined and must devote the time for the study. The investigator must be fully aware of the proper use of the investigational product and he/she should be thoroughly familiar with the study protocol.

5. Procedures

All clinical trial procedures should be consistent with GCP/applicable regulatory requirements. The clinical trial information generated from studies should be recorded, handled and stored in a way that allows for accurate reporting, interpretation and verification. The case report form (CRF) is usually prepared by the sponsor of the study. The investigator should ensure that the pages in the CRF are completed legibly, at every visit. The recording of data in the CRF must be systematically entered to allow proper verification. There are strict rules for amending data in CRFs. The recording of data has to be done in accordance with GCP requirements. It may be tedious process, but necessary to ensure integrity of data. ICH-GCP guidelines stipulate that the study monitor should have direct access to all subject files, in order to ensure that all necessary data that have been recorded is correct.

It is the responsibility of the investigator to obtain information from the subjects about any adverse events that may have occurred during study, as this information may not be readily offered by the subject unless, specifically asked for. An adverse event is any untoward medical occurrence in a trial subject who has been administered a pharmaceutical product which does not necessarily have a causal relationship with the treatment. All such adverse events during the course of study must be properly recorded. The serious adverse events (as defined by study protocol) must be reported immediately to the sponsor and the Ethics Committee.

6. The Investigational Product

The products under investigation should be manufactured, handled and stored in accordance with applicable Good Manufacturing Practice (GMP). They should be used in accordance with the approved protocol.

The responsibilities of all the stakeholders involved in the clinical study, such as the sponsor, the monitor, the investigator and the patient must be fully understood. The cost of individual clinical trials has escalated substantially since the implementation of GCP guidelines. The clinicians who undertake clinical trials must be fully aware of the importance of conducting the study efficiently and in the shortest time. It is, therefore, necessary that the investigator reads the protocol carefully before agreeing to undertake the study. He/she should decide whether it is possible to recruit subjects based on the inclusion and exclusion criteria and whether, the study procedures fit into his/her usual clinical practice. He/she must regularly check the performance against the agreed recruitment rate and identify problems, as they occur and not hibernate them until it becomes unmanageable. For successful implementation of the study, the investigator needs a committed team of workers including, a good research assistant and efficient co-investigators, all of whom should be very familiar with GCP and the study protocol. The meetings of the members of the clinical team should be held regularly for study-oriented discussions and the minutes of such interactive meetings should be recorded.

Independent Ethics Committee (IEC)

The basic responsibilities of IEC are as follows.

1. To ensure competent reviews of all ethical issues related to the project proposal submitted.
2. To ensure protection of dignity, rights and well beings of the potential participants in clinical trials.
3. To ensure the expression of universal ethical values and international scientific standards in terms of local community values and customs.
4. To facilitate development and the education of a research community that is responsive to local health care requirements.

Composition of IEC

The IEC is represented by around eight members drawn from multidisciplines and multisectors. There is no specific recommendation for maximum number, but it is presumed that large size of the Committee will make it difficult in reaching consensus. Five members are sufficient to form the quorum. The Chairperson of the Committee should be an outsider to maintain the independence in functioning. He/she should be a qualified and experience person. The Member Secretary generally belongs to the same institution and he/she conducts business of the Committee.

The following is the composition of IEC.

1. Chairperson of IEC
2. Basic medical scientists 1-2, of which preferably one should be a pharmacologist.

3. One or two clinicians from different institutes
4. One retired judge or any other legal expert
5. One social scientist or representative of NGO
6. One philosopher/theologian/ethicist
7. One common person
8. Member Secretary

There should be adequate representation of age, gender, community, etc. on IEC. The Terms of References should include procedure for conduct, removal, replacement and resignation.

Basic Principles of GCP – A Summary

1. The clinical trials must be conducted in accordance with the ethical principles that have their origin in the Declaration of Helsinki. It should be consistent with GCP and the applicable regulatory requirement(s).
2. Before a clinical trial is undertaken, foreseeable risks and inconveniences should be weighed against the anticipated benefit for the individual trial subject and society.
3. A trial should be initiated and continued only, if the anticipated benefits justify the risks.
4. The available non-clinical and clinical information on an investigational product should be adequate to support the proposed clinical trial
5. The rights, safety, and well-being of the trial subjects should prevail over interests of science and society.
6. Each individual involved in conducting a trial should be qualified by education, training, and experience to perform his/her respective task.
7. Clinical trial must be scientifically sound. A clear and detailed protocol be described for the same.
8. Freely given informed consent should be obtained from every subject prior to participation in clinical trial.
9. A trial should be conducted in compliance with the protocol that is approved by Institutional Review Board (IRB)/Independent Ethics Committee (IEC).

10. The medical care given to, and medical decisions taken on behalf of the subjects should always be the responsibility of a qualified physician/qualified dentist as applicable.
11. Investigational product should be manufactured, handled, and stored in accordance with applicable good manufacturing practice (GMP). It should be used in accordance with the approved protocol.
12. Systems with procedures that assure the quality of every aspect of the trial should be effectively implemented.
13. All information pertaining to clinical trial should be recorded, handled, and stored in a way that allows its accurate reporting, interpretation and verification.
14. The confidentiality of records that could identify subjects should be protected. It is necessary to respect the privacy and confidentiality of the subject in accordance with the applicable regulatory requirement(s).

CHAPTER 10

The Narcotic Drugs and Psychotropic Substances (NDPS) Act, 1985 and Rules, 1985

The developments in science and technology have certainly contributed to improvized global health care and the life span of the human beings world wide has seen upward progression. With the discovery of new drugs and formulations, there has been appreciable change in our understanding and treatment of diseases. The technological developments, on the other hand, have been misused for production of powerful psychotropic substances and in modification of narcotic drugs, which are being misused by a considerable segment of global population. The earlier provisions of the Acts, including the Dangerous Drugs Act, 1930, were inadequate to prohibit illicit trade of NDPS. These drugs of abuse have played havoc with the society. It was to counter procurement, manufacture and trade of such harmful substances, a new Act was enacted, keeping in view the recommendations of various international conventions. Today, India adopts global strategy to counter ill-effects of NDPS.

Objective

It is the Act to consolidate and amend the law relating to narcotic drugs; to make stringent provisions for the control and regulation of operations relating to narcotic drugs and psychotropic substances; to provide for the forfeiture (seizure) of property derived from or used in illicit traffic in narcotic drugs and psychotropic substances; to implement the recommendations of the international conventions on narcotic drugs and psychotropic substances; and for matters connected therewith. The Act extends to the whole of India.

The Act is divided into six Chapters and it has 83 Sections.

Chapter I is preliminary,

Chapter II deals with Authorities and officers,

Chapter II-A deals with National Fund for Control of Drug Abuse,

Chapter III deals with Prohibition, Control and Regulation of NDPS,

Chapter IV refers to Offences and Penalties,

Chapter V deals with Procedure,

Chapter V-A pertains to Forfeiture of Property derived from or used in Illicit Trafficking,

Chapter VI is miscellaneous.

Definitions

1. Cannabis (Indian Hemp)

It contains leaves and flowering tops of pistillate plants of *Cannabis sativa (C.indica)* : Family—Cannabinaceae.

2. Cannabis Products

(i) **Charas** is the crude or purified resin obtained from the cannabis plant and includes concentrated preparation and resin known as Hashish oil.

(ii) **Ganja** is from the flowering or fruiting tops of the cannabis plant (excluding the seeds and leaves not accompanied by the tops), by whatever name they may be known or designated.

(iii) Any mixture with or without any neutral material of any of the above forms of cannabis or any drink prepared therefrom.

3. Coca-derivatives

It includes:

(i) Crude cocaine as the extract of coca leaf which can be used directly or indirectly for the manufacture of cocaine.

(ii) Ecgonine and all the derivatives of ecgonine from which it can be recovered.

(iii) Cocaine-methyl ester of benzoylecgonine and its salts.

(iv) All preparations containing more than 0.1% of cocaine.

4. Coca Leaf

(i) It is the leaf of the coca plant *Erythroxylon* except the leaf from which all ecgonine, cocaine and any other ecgonine alkaloids have been removed.

(ii) Any mixture thereof with or without any neutral material but does not include any preparation containing not more than 0.1% of cocaine.

5. Coca Plant

It is the plant of any species of the Genus *Erythroxylon* i.e. E. coca, *E. truxillence* and other species of *Erythroxylon*.

6. Illicit Traffic

In relation to narcotic drugs and psychotropic substances, it means:

(i) Cultivating any coca plant or gathering any portion of coca plant,

(ii) Cultivating the opium poppy or cannabis plant,

(iii) Engaging in the production, manufacture, possession, sale, purchase, transportation, warehousing, concealment, use or consumption, import inter-State, export inter-State, import into India, export from India or transhipment of narcotic drugs or psychotropic substances,

(iv) Dealing in any activities in narcotic drugs or psychotropic substances other than those referred to under sub-clauses (i) to (iii) as above,

(v) Handling or letting out any premises for carrying on any of the activities referred to in sub-clauses (a) to (d) given above.

7. Controlled Substances

Any substance which the Central Government may, having regard to the available information as to its possible use in the production or manufacturing of the narcotic drugs or psychotropic substances or to the provisions of any international convention by notification in Official Gazette, declare to be a controlled substance.

8. Manufactured Drug

(i) It means all coca derivatives, medicinal cannabis, opium derivatives and poppy straw concentrate,

(ii) Any other narcotic substance or preparation which the Central Government may having regard to the available information as to its nature or to a decision if any under any international convention by notification in Official Gazette declare to be a manufactured drug but, does not include any narcotic substance or preparation which the Central Government may, having regard to the available information as to its nature by notification in Official Gazette declare not to be a manufactured drug.

9. Opium

It means:

(i) Medicinal opium—the opium in powder, granulated or any other form which has undergone the processes necessary to adopt it for medicinal use in accordance with the requirements of Indian Pharmacopoeia or any other Pharmacopoeia,

(ii) Prepared opium—any product of opium obtained by a series of operations designed to transform opium into any extract suitable for smoking,

(iii) Phenanthrene alkaloids, namely, morphine, thebaine, codeine and their salts,

(iv) Diacetyl morphine i.e., the alkaloid known as diamorphine or heroine and its salts, and

(v) All prepatations containing more than 0.2% of morphine or containing any diacetyl morphine.

10. Opium Poppy

It is the plant of the species of *Papaver somniferum* L.(Papaveraceae) and the plant of any other species of *Papaver* from which opium or any phenanthrene alkaloid can be extracted and which the Central Government may by notification in the Official Gazette declare to be opium poppy for the purposes of this Act.

11. Poppy Straw

It represents all parts except the seeds of the opium poppy after harvesting, whether in their original form or crushed or powdered and whether or not juice has been extracted therefrom.

12. Poppy Straw Concentrate

It means the material arising when poppy straw has entered into a process for the concentration of its alkaloids.

13. Psychotropic Substance

It is a natural or synthetic substance or its salt or preparation or any natural material included in the list of psychotropic substances specified in the Schedule to the Act.

The list of some of the psychotropic substances covered under the Schedule to the Act including, tetrahydrocannabinol, amphetamine, etycyclidine, rolicyclidine, psilocybine, tenocyclidine, dexamphetamine, camazepam, clonazepam, clotizepam, diazepam, cloxazolam, ketozolam, alprazolam, estazolam, bromazepam, haloxazolam, amobarbital, pentobarbital, secobarbital, barbital, allobarbital, LSD, LSD-25, mesocarb, cathine, tetrazepam, triazolam, prazepam, oxazepam, loprazolam, lorazepam, mazindol, meprobamate and others.

Authorities and Officers

The Central Government has to:

1. Fulfil obligations under international conventions,
2. Coordinate with international agencies,
3. Coordinate actions of Central Government and State Government Officers,
4. Extend assistance to foreign authorities when asked for,
5. Undertake identification of problem, treatment of addicts, education against drug abuse, after-care of de-addicted persons, rehabilitation and social registration of de-addicted persons and
6. Any other matters.

Officer of Central Government is Narcotic Officer of India and other officers for opium poppy cultivation, manufacturing, processing, storage, distribution, transport, import-export and all other activities.

Officers of State Government are designated as the Government's kingsfit for the Act. They coordinate with officers of the Central Government and also act independently in regard to all offences pertaining to cannabis and products, other NDPS.

Narcotic Drugs and Psychotropic Substances Consultative Committee

The Central Government appoints this advisory committee to advise Central Government and State Governments on the matters pertaining to NDPS Act, 1985. The Committee consists of the Chairman and not more than 20 persons appointed by the Central Government. The tenure, constitution of sub-committees, cooption of members, filling up of casual vacancies, allowances, etc., are workout by the Government in consultation with this Committee.

National Fund for Control of Drug Abuse

The Central Government by notification in Official Gazette creates National Fund for implementing various measures to control the drug abuse. The money generated is from:

1. The grant approved by the Parliament.
2. Sale proceeds of property confiscated or forfeited during the raids by the officers.
3. Financial support from individuals or institutes which is usually, income-tax free.
4. Any income from investments of amounts credited.

The fund is to be used for preventing or combating illicit traffic of NDPS, controlling drug abuse, educating masses, rehabilitation of drug addicts, etc. The Central Government constitutes the Governing Body with Chairman who is not below the rank of additional secretary in Central Government, and not more than six members.

The Annual Report covering the activities of the Governing Body and utilization of the national fund is submitted to the Central Government and placed before the Parliament.

Prohibition Control and Regulation of NDPS

No person without the permission of the Central Government shall:

1. Cultivate coca plant or gather any portion of coca plant,
2. Cultivate opium poppy or cannabis,
3. Produce, manufacture, possess, sell, purchase, transport, warehouse, consume, use, import and export any NDPS except for medical and scientific purpose with Government appoval.

The Central Government may fix time frame for cultivation of opium poppy, frame regulations for delivery of opium by cultivators, fix prices to be paid to cultivator, issue permits for manufacture, possession, export, import and prescribe other conditions for regulation. The Central Government appoints officers for this purpose.

Powers of State Government

The State Government may permit, control and regulate all activities pertaining to poppy straw, cannabis excluding charas and only possession, transport, sale, purchase, import and export of manufacturing drugs other than opium and coca leaves. The State Government is empowered to declare a place as warehouse for poppy straw, define limits of licence for cannabis, fix price to be paid for cultivation of cannabis and impose other conditions of licence and permit.

No external dealings of NDPS are permitted. There are special provisions for coca plant and leaves which do not contain alkaloid cocaine and used as a flavouring agent. Special orders of State Government are required for cannabis cultivation for obtaining fibres for industrial use or seeds for horticultural purpose.

Offences and Penalties

1. For offence in relation to poppy straw, coca plant and leaves, prepared opium poppy or cannabis plant, there is punishment of rigorous imprisonment of 10-20 years and fine of not less than Rupees ₹ 1 lakh in case of first conviction. However, the subsequent conviction is punishable with rigorous imprisonment of 15-30 years and fine of ₹1.5 to 3 lakhs.

2. For contravention or offence relating to ganja, punishment is up to five years and fine up to ₹ 50,000 or both.

3. In relation to manufactured drugs and preparations, psychotropic substances, external dealings of NDPS etc., for first conviction, the punishment is rigorous imprisonment of 10-20 years and fine of ₹ 1-2 lakh. The subsequent conviction is punishable with rigorous imprisonment of 15-30 years and fine of ₹ 1.5-3 lakhs.

4. For allowing premises to be used for offence, first conviction is punishable with rigorous imprisonment of 10-20 years and fine of ₹ 1-2 lakhs, while for subsequent conviction, the punishment is rigorous imprisonment of 15-30 years and fine of ₹ 1.5-3 lakhs.

5. The offence for illegal possession of NDPS by an individual in small quantity for personal consumption is punishable with six months to one year imprisonment or fine or both.

6. For illegal traffic of NDPS and also for harbouring offenders or helping offenders, the punishment is imprisonment of 10-20 years and fine of ₹ 1-2 lakhs. For subsequent conviction, rigorous imprisonment is for 15-30 years and fine of ₹ 1.5-3 lakhs.

7. **Death penalty :** There is a provision of death penalty for certain serious offences that are committed after previous conviction. If any individual or firm is found to possess more quantities of NDPS as mentioned in Table 9.1 without permission, license, etc., and if the offence is repeated, the death penalty may be awarded.

Death penalty may also be awarded for a serious offence of financing directly or indirectly such a big crime as indicated. Special Courts may be constituted with single judge appointed by the Chief Justice of the High Court to dispose of the cases of NDPS.

Issue of Warrant and Authorization

The warrant is issued by Metropolitan Magistrate or Magistrate of I class or II class specially empowered by the State Government for arrest of person or persons whom he has reason to believe has violated provisions of the Act and also warrant for search of any building or premises or and other articles during day or night.

An officer of gazetted rank in Central Government from the departments of central excise, narcotics, customs, revenue intelligence, border security force or army or any other department of Central Government shall be authorized for the purpose of this Act. The State

Table 10.1

NDPS	Quantity (attracting death penalty)
Narcotics	
Opium	more than 10 kg
Morphine	more than 1 kg
Heroin	more than 1 kg
Codeine	more than 1 kg
Thebaine	more than 1kg
Cocaine	more than 500 gm
Hashish	more than 20 kg
Any mixture of above drugs more than 1.5 Kg.	
Psychotropic Substances	
LSD	500 gm
Tetrahydrocannabinol (from cannabis) (THC)	500 gm
Amphetamine	1.5 kg
Methamphetamine	1.5 kg
Methaqualone	1.5 kg
Mixture of any of the above psychotropic substances	1.5 kg

Government may appoint officer of revenue, drug control, excise, police or any other department as the officer authorized and empowered to implement the warrant. The officer may depute his subordinate but superior to the rank of peon, sepoy or constable to search premises or conveyance or transport and arrest a person provided he has reason to believe that the provisions of the Act are violated.

Any addict who is not charged with any offence earlier and seeks to undergo medical treatment for detoxification of deaddiction from an institution or hospital recognized by Government or Local Authority and undergoes such a treatment shall not be liable to prosecution only once in his/her lifetime.

Disposal of seized NDPS is done as per the procedure laid down by the Central Government. The obligation of officers of the Central and State Government is to assist each other for the effective implementation of the Act.

Any authorized officer can enter into and search building, conveyance or transport of place and in case of resistance, break open the lock and door, remove any obstacle to such entry, seize NDPS and other materials, conveyance, animal used in illicit traffic, detain any person for interrogation, provided he/she believes that a search warrant or authorization cannot be obtained for want of time and if action is delayed, the offender may escape or NDPS may disappear. This action can be taken only after sunset and before sunrise, provided the officer has authentic information of the crime or a complaint in writing. If this authority is misused by the officer and proved so, he/she is punishable for imprisonment of six months or fine of ₹ 1000/- or both.

For giving false information about the offence in relation to NDPS to the officer concerned, an individual may be punished with two years imprisonment or fine or both.

The officer refusing to perform duty or showing negligence in discharge of his duties, imprisonment of one year or fine or both are possible.

Chapter 11

Medicinal and Toilet Preparations (Excise Duties) Act, 1955 and Rules, 1956

Alcohol is an important industrial solvent, excellent preservative, and a therapeutic agent. It is, however, likely to be misused and it could be a drug of addiction (euphoric drink). Before enactment of this Act, there was chaotic condition prevailing in the country in relation to price structure of alcohol as raw material and formulations containing alcohol. Some States were rich in production of alcohol, while others were utilizing this as raw material in manufacture of formulations. In the absence of uniform excise policy, the price structure of alcohol containing medicines was varying from State to State within the country. In order to overcome this difficulty and to ensure that uniform structure of excise duty for alcohol products exists, this Act was enacted.

It is an Act with provision for levy and collection of excise duties on medicinal and toilet preparations containing alcohol, opium, Indian hemp (cannabis) or other narcotic drugs. The Act is effectively implemented throughout India from 1st April, 1957. The Act has eleven Chapters and 21 Sections.

Important Definitions

1. **Narcotic:** It means coca leaf and its derivatives or opium and its derivatives or Indian hemp and any other substance capable of causing or producing in human being dependence, tolerance and withdrawal syndromes.

2. **Toilet Preparation:** It means any preparation used in the toilet, the human body or in perfuming apparel of any description. It also covers substances used to cleanse, improve or alter the complexion, teeth, skin or hair and includes perfumes and deodorants.

3. **Absolute Alcohol:** It is alcohol conforming to the British Pharmacopoeial specification for dehydrated alcohol.

4. **Bonded Manufactory or Laboratory:** It is the premises approved and licensed for the manufacturing and storage of medicinal and toilet preparations containing alcohol, opium, Indian hemp and other narcotic drugs on which duty has not been paid.

Non-bonded Manufactory or Laboratory

It is the premises approved and licensed for manufacture and storage of preparations containing alcohol, opium and Indian hemp and other narcotic drugs on which duty has been paid.

Denatured Spirit or Denatured Alcohol

It means alcohol of any strength which has been rendered unfit for human consumption by addition of substances, approved by Central Government or by State Government with the approval of Central Government.

Excise Commissioner

He /She is the Head of the Excise Commission of a State and includes a Prohibition Commissioner or an officer designated in any State for the purpose.

Finished Store

It is portion of non-bonded or bonded laboratory which is set apart for the storage of finished preparations.

Spirit Store

It means that portion of bonded or non-bonded laboratory which is set apart for storage of alcohol, opium, Indian hemp and other narcotics or narcotic drugs purchased free of duty or at prescribed rates of duty specified in the Schedule to the Act.

[The definitions of coca derivatives, coca leaf, opium and opium derivatives and cannabis (Indian Hemp) are as given under the Narcotic Drugs and Psychotropic Substances Act, 1985 under Section 6 of the Act.]

The Central Government may, by notification in the Official Gazette, provide that from such date as may be specified, no person shall engage in the production or manufacture of any dutiable goods or their ingredients or specified container or labels for such goods, except under the authority and in accordance with the terms and conditions of a licence granted under this Act. Every licence shall be granted for specific area and period subject to prescribed conditions and restrictions.

No licence for alcoholic medicinals is given unless there is a licence issued under Drugs and Cosmetics Act (DCA). A licensee (under DCA) can only approach the Excise Commissioner for issue of alcohol, if the formulations containing alcohol are being manufactured. The licence on payment of specified fee can be obtained subject to fulfillment of conditions for manufacture in bond and outside bond of medicinal and toilet preparation, ayurvedic, homoepathic and unani medicines, as well as for bonded warehouse.

This Act covers manufacturing in bond, manufacturing outside bond, manufacturing of homoepathic, unani and ayurvedic preparations containing alcohol, issue of preparations from bonded laboratory, and their inter-State movements and responsibilities of the officers.

Section 3 of the Act stipulates the duties of excise to be paid and collected at the rates specified in the Schedule on all dutiable goods manufactured in India both of bond and non-bond categories.

Manufacturing in Bond (Bonded Manufactory)

The licence is issued from office of the Excise Commissioner of the State for alcohol. Application should be made two months in advance prior to date of manufacturing along with requisite fee as indicated.

The application should include:

1. Name, addresses, and site of bonded laboratory if it is a firm name and address of every partner of the firm; if it is a company, its registered name, address of the Managing Director and Directors.

2. Capital to be invested for bonded preparation.
3. Number and description of apparatus, machinery, still and maximum amount of alcohol to remain at one time.
4. Approximate date of starting manufacturing; requirement of excise officer, either whole-time or part-time.
5. List of preparations containing alcohol; licence number issued by Licensing Authority under DCA.
6. Elevation plan, plan of different rooms, details of doors and windows, and quarters for excise staff are to be provided.
7. In case of firm, partnership deal and in case of companies, list of association and latest balance sheet of the company.

The Licensing Authority shall enquire into the purpose of manufacturing of such formulations containing alcohol and narcotic drugs; qualifications of the staff employed, previous experience of technical personnel; suitability of the proposal; and financial position of the applicant. The Licensing Authority shall also look into the equipment, requirement of alcohol and fix an amount of security. After satisfying all conditions, the Licensing Authority issues licence for one year relating to the premises approved after inspection and for manufacture of product containing alcohol or narcotic drug. Separate licence is issued for each premises and the licence is not transferable.

The licence issued should be displayed properly in the premises. Licensing Authority may amend or alter the conditions of licence. Renewal of licence should be done one month before its expiry or else 25 percent extra fee is charged. Conviction of violation of provision of the Act is possible under sections 109 or 116 of Indian Penal Code. Before canceling or suspending the licence, sufficient opportunity is given to the licensee to explain his/her position.

(i) Structures of the Bonded Manufactory

A separate plain spirit store is required to be provided in the bonded premises. There should be only one entrance to the laboratory, only one door for each of its compartment, a separate room for manufacturing of medicinal preparations, one or more rooms for storage, a separate room for manufacturing of toilet preparations and their storage, and a room for excise officer incharge near the entrance with furniture. The bonded laboratory should be opened and closed in the presence of excise officer incharge and secured with excise ticket locks. Every window in the bonded laboratory should be provided with iron rods of not less than 19 mm in thickness set apart at a distance of not more than 102 mm. The iron rods should be embodied in brick construction to a depth of 51 mm at each end. The windows should be covered from inside with strong wire netting, the aperture of which should not exceed 25 mm in diameter. Each room in bonded laboratory should have a board indicating name and a serial number of the room and the purpose for which it is used, painted in oil colour. The pipes from sinks in the laboratory should be connected to closed drainage, ultimately forming part of general drainage system. The gas and electric supply should be arranged in such a way that they are cut off at the end of day's work from one place. All the regulators or switches should be securely locked. Permanent vessels in bonded laboratory should bear the serial number and statement of full capacity of the vessel.

The spirit should be obtained from the spirit warehouse approved by the Excise Commissioner. The indent should be sent in duplicate in prescribed form counter-signed by the officer incharge of the laboratory. Alcohol is issued in duly sealed containers and under intimation to the Excise Officer concerned. No wastage during transhipment is permissible and for any loss due to negligence of the manufacturer, excise duty has to be paid.

(ii) Verification and Storage of Alcohol

On its arrival in bonded laboratory, alcohol is measured in volume and strength. Entries are made in register and stored in spirit store room under excise lock with perfect coordination between excise officer incharge and officer of manufacturing unit. The alcohol is issued from spirit store from time to time in accordance with the procedure laid down under Rules.

(iii) Issue of Alcohol from Store

The manufacturer should calculate requirement of alcohol and hand it over to the excise officer. All ingredients should be kept ready and on receipt of alcohol, the solvent should be mixed immediately in presence of excise officer. The percolators or other vessels containing alcohol during the process of manufacturing and storage should be labeled with the name of the product, batch number, description, quantity of alcohol used, date of manufacture and quantity of preparation removed. The preparation should be immediately removed to the store, measured and stored in vessels, provided it is entered in register and given a batch number. Up to 200 ml from each batch is permitted to be withdrawn without payment of duty for determination of alcoholic strength of preparation. A separate account of amount of samples used by the manufacturer for the purpose of analysis is maintained. Any amount of preparation left after analysis should be returned and mixed with the main batch. All entries must be initialed by the excise officer incharge. The excise officer may also withdraw two samples from each batch (not more than 150 ml) for sending one sample to the chemical examiner and retaining the other sample until the report is received. If the report of the chemical examiner indicates that the strength of alcohol in the preparation is more than the one declared by the manufacturer, the strength reported by the chemical examiner is entered in stock book and duty for excess strength of alcohol has to be paid by the manufacturer. Any preparation wherein strength beyond tertiary proof is established, it is either destroyed, reprocessed or disposed of as per the instructions of Excise Commissioner. The Excise Commissioner may exempt payment of duty, if convinced of any accidental loss of alcohol, except theft during the manufacturing or storage. The wastage limit for alcohol is fixed by the Central Government or State Government.

Bulk storage of preparations is done in jars or bottle of not less than 2.25 litres in volume. The container is labeled with batch number, content, strength of alcohol, etc. This preparation should be distributed in containers of not less than 50 ml capacity. However, in special cases, Excise Commissioner may authorize use of small containers. The containers with all necessary particulars are arranged in racks in the Bonded Laboratory. Any deficiency detected should be brought to the notice of excise officer who shall keep record and report to Excise Commissioner. The Excise Commissioner may or may not ask manufacturer to pay duty for such deficiencies, depending upon the merit of the case.

(iv) Issue from Bonded Laboratories

The manufacturer makes application to excise officer and pays excise duty or it can be adjusted against any advance duty paid by him. Only on payment of calculated duty, the goods containing alcohol and narcotic drugs are permitted to be taken out of bonded laboratory by the excise officer.

Exemptions: The preparations supplied to Government hospitals, dispensaries and charitable hospitals and institutions, which supply medicines to poor as certified by the district medical officer, are exempted from payment of excise duties.

Penalties: For contravention of provisions of the Act or evasion of excise duties or failure to supply required information, the punishment is imprisonment for six months with or without fine. The Court may order seizure of dutiable goods and hand them over to the Government. The owner of the land should inform the magistrate, excise officer or police or land revenue officer of any illegal manufacturing taking place in his/her place, otherwise, he/she is liable for imprisonment up to six months with or without fine.

Any officer without reasonable ground detaining the goods with an intention of harassment or torture may be fined up to rupees two thousand.

Any officer who refuses to perform duties unless he has obtained permission to withdraw shall be punishable with three months imprisonment with or without fine.

The Central Government may order that from a specified date no person shall engage in production of dutiable goods or part thereof.

Any officer authorized by the Excise Commissioner shall have free access at all reasonable times to any licensed premises for a thorough inspection. If obstructed to discharge of duties of the officer, there is provision for penalty of rupees five hundred. Any authorized excise officer may arrest any person whom he/she has reason to believe that the person has violated provisions of the Act and is liable to be punished under this Act.

Ayurvedic Preparations and Homoeopathic Preparations

Ayurvedic preparations are of two types:

1. Self-generated alcohol containing preparations like Asavas and Aristas
2. Prepared by distillation or addition of alcohol.

   Self-generating preparation containing alcohol were earlier exempted from excise duty. However, now all such preparations are chargeable with four percent *ad valorem* excise duty. Those preparations with more than two percent self-generated alcohol and being capable of consumed as a beverage are charged duty of four percent *ad valorem*.

Ayurvedic practitioners are exempted for licence from excise authorities if the samples are being used for their own patients. However, the excise officer should be allowed to take the sample, if required.

For homoeopathic preparations capable of consumption as ordinary alcoholic beverages, the excise duty of four percent *ad valorem* is required to be paid.

Manufacturing Outside Bond (Non-bonded Manufactory)

The manufacturing and sale operations are restricted only between sunrise and sunset and days and hours as fixed by the Excise Commissioner.

Essential requirements of manufactory are:

1. Separate laboratory spirit store and finished store.
2. Only one entrance to non-bonded manufactory and one door each to laboratory, spirit store and finished store.

3. The iron bars in windows should be not less than 19 mm in thickness, set not more than 102 mm apart and fixed in brick construction to the depth of not less than 51 mm. The aperture of wire netting to the window should not exceed 25 mm in diameter.

4. A separate spirit store for rectified spirit purchased at the specified duty.

5. All pipes from wash basins and sinks are connected to closed drainage forming part of general drainage system.

6. All electric and gas connections should be well secured with one cut-off switch.

The State Government may relax conditions of requirements of infrastructure if alcohol consumption is less. Alcohol is procured by the manufacturer after payment of excise duty. The form for payment of duty is filled up in triplicate, one copy is to be given to spirit warehouse keeper, second to officer incharge and third to be retained by the licensee. The duty for alcohol purchased is to be paid to Government treasury and challan enclosed. Thereafter, the treasury officer sends an advice to officer incharge. Sealed containers are required to be used for procurement and storage of alcohol. The alcohol purchased cannot be resold.

Manufacturing, storage and sale of preparations should be carried out only in licensed premises. Batch number should be given and separate finished store is provided for alcoholic products. All necessary entries of operations carried out are maintained in stock register and the same should be made available for inspection by the excise officer and drug control authority. Excise officer may take up to ten percent of the preparation for analysis and the same is sent to the chemical examiner. If the chemical examiner of government gives the report that alcohol content is more than what is claimed, the penalty charged is ten times or maximum of ₹ 2000/- per batch. If this is frequently done by manufacturer, licence may be withdrawn.

The Schedule (Section 3 of the Act)

Duties of Excise to be levied and collected on Certain Goods

S.No	Description of Medicine	Rate of Duty
I	Allopathic medicinal Preparations :	
[A]	Medicinal preparations containing alcohol capable of being consumed as ordinary alcoholic beverages,	
	(i) For patent and proprietary medicines *ad valorem*	20%
	(ii) For others	20% *ad valorem*
[B]	Medicinal preparations containing alcohol capable of being consumed as ordinary alcoholic beverages	
	(i) Medicinal preparations containing known active ingredients in therapeutic quantities.	20% *ad valorem*
	(ii) For others	20% *ad valorem*
[C]	Medicinal preparations without alcohol but containing narcotics or narcotic drug.	20% *ad valorem*
II	Ayurvedic Unani or other indigenous system of medicines	
[A]	Medicinal preparations containing self-generated alcohol not capable of being consumed as ordinary alcoholic beverages	4% *ad valorem*
[B]	Self-generated alcoholic medicinal preparations capable of being consumed as ordinary alcoholic beverages	4% *ad valorem*
[C]	All other preparations containing alcohol with 6% are prepared by distillation or to which alcohol has been added	*ad valorem*
[D]	Medicinal preparations without alcohol but with 20% narcotic drug or narcotic	*ad valorem*
III	Homoeopathic preparations containing alcohol	4% *ad valorem*
IV	Toilet preparations containing alcohol or a narcotic drug or narcotic	50% *ad valorem*

CHAPTER 12

The Industries (Development and Regulations) Act, 1952

This Act was enacted on 8th May 1952 and is applicable to whole of India. The Act is aimed at monitoring, regulating and developing the industrial activities in the country. The industrial units covered under the provisions of this Act also include the manufacturing units of pharmaceutical machinery, medical and surgical appliances, drugs and pharmaceuticals, cosmetics and soap perfumery and toiletary items, diagnostic kits and related fields.

The salient features of this Act are :

1. Central Advisory Council (CAC)

It is constituted by the Central Government and comprises Chairman and nominated members, not more than 30. It advises Central Government on matters arising out of implementation of the provisions of this Act. Central Advisory Council is an advisory and policy-making body assisting Central Government and State Governments and Union Territories on the technical, as well as, professional matters pertaining to the Act. The members of Central Advisory Council include:

(i) Representatives of manufacturers.

(ii) Representatives of employees in such manufacturing units.

(iii) Representatives of consumers of the products manufactured by such units.

(iv) Others including subject experts, government officials, etc.

2. Development Council (DC)

It is also constituted by the Central Government comprising Chairman and members drawn from categories of manufacturers, employees, consumers, experts and government officials. There is one Development Council for the scheduled type of industry. The term of office of Development Council, procedure to be followed for meetings, filling up of vacancies and other operational matters are as per the instructions of Central Government. Development Council is accountable to the Central Government for its performance and it is charged with following responsibilities:

(i) Survey for production in a particular sector,

(ii) Planning the strategies for production,

(iii) Deciding optimum and minimum production of commodities,

(iv) Developing service structure and planning for economic viability,

(v) Scrutinizing carefully the market production and regulating the price line,

(vi) Planning strategy for making available to manufacturers raw materials required for production.

Development Council is a corporate body. It can sue or be sued. It has to submit annual report to Central Government which is tabled in the Parliament. The Central Government may levy a duty on manufacturers and make available such funds for various activities of Development Council. The Development Council is also required to encourage research and development in the specific areas of priority by way of making financial assistance available for such research efforts.

3. Regulation of Schedule Industries

The registration of all industries is required to be completed within six months of enactment of the Act. No industrial unit can exist without registration or licence after this deadline. No starting of units in private sector is permitted without licence. New licence for new articles to be manufactured is required to be obtained. The State Government may start a unit without licence after obtaining the permission of the Central Government. The Central Government through this Act is fully empowered to make rules and frame guidelines on the advice of Central Advisory Council. The Central Government may investigate for fall in volume of production, regulation of production, deviation of price line, and irregularities in distribution, and in public interest issue the guidelines to the industrial units to be complied with within stipulated time. If the Industrial unit ignores the notice issued by Central Government and does not comply with stipulated conditions, the Central Government may authorize person/persons to take over management of the unit partly or wholly initially for five years and thereafter two years extension is possible six times (Total take over - $5 + 2 \times 6 = 17$ years). The investigation team appointed by Central Government has powers of civil court which works according to section 21 of Code of Criminal Procedure, 1898. The investigation team records, witnesses, interrogates the persons concerned and submits the report to the Central Government.

Powers of Inspection Officer

The Central Government identifies an officer and authorizes him to ensure the accomplishment of the task of implementing the provisions of this Act. The officer may enter the industrial premises and carry out the scrutiny of all relevant records, conduct enquiry with persons working in industrial units and report back to Central Government. The Central Government may delegate these powers in consultation with State Government to the officers of State Government.

Penalties

1. For false or misleading statement or information, imprisonment is possible for three months or fine up to ₹ 2000 or both.
2. If person fails to comply with the instructions of Central Government, starts a unit without license or works in contravention with provisions of the Act, the penalty is imprisonment up to six months or fine up to ₹ 500 or both for first conviction and for subsequent conviction, in addition to initial penalty, a fine of ₹ 500 every day.

Chapter 13

Medical Termination of Pregnancy Act, 1971 & Rules, 2003

This Act provides for termination of certain pregnancies by registered medical practitioners. It also covers all related matters associated therewith. It extends to whole of India except J & K and came into force w.e.f. 1-4-1972.

Definitions

1. **Guardian** is a person taking care of a minor or mentally ill person.
2. **Registered Medical Practitioner (RMP)** is a person with recognized medical qualification as under clause (h) of Section 2 of the Indian Medical Council Act, 1956 whose name is entered in State Medical Register and who possesses such experience or training in obstetrics and gynecology as prescribed under Rules to the Act.

Termination of Pregnancies and Place for Termination

The RMP is permitted to undertake termination of pregnancy upto twelve weeks, if he / she is of the opinion that the continuance of pregnancy would involve a risk to the life of the pregnant woman or cause grave injury to her physical or mental health or to the child on birth. For the length of pregnancy between twelve and twenty weeks, the opinion of not less than two registered medical practioners is mandatory for termination of pregnancy.

However, in case of emergency situation, the RMP is permitted to undertake termination of pregnancy to save the life of the pregnant woman.

The pregnancy may be terminated at Government hospital or a place approved for the purpose of this Act by Government or a District Level Committee constituted by the Government with chief Medical Officer as the chairperson. The strength of members of such Committee shall be between three and five including, Chairperson.

Punishment for Offence

An unauthorized person involved in termination of pregnancy is punishable with two to seven years of rigorous imprisonment. Whoever terminates any pregnancy at a place not approved for the purpose or the owner of such a place is also punishable with two to seven years of rigorous imprisonment.

Powers to Make Rules and Regulations

The Central Government may with the approval of Parliament by notification in the Official Gazette, make the rules to carry the provisions of the Act. The State Government is empowered to make Regulations, from time to time, to facilitate the process of the authorized termination of pregnancy by registered medical practitioner within the framework of the Act and without diluting its provisions. Anything that is done by RMP in good faith and if it is prooved so, shall not attract any legal action.

Medical Termination of Pregnancy Rules, 2003

The Rules framed by the Central Government have come into force from the date of their publication in Official Gazette. These Rules cover the details regarding the composition and tenure of the District Level Committee, experience of the registered medical practitioner permitted to undertake termination of pregnancy and approval of place for termination of pregnancy.

1. ***The District Level Committee***: It is for approval of the place for termination of pregnancy. It shall be represented by a gynecologist / surgeon / anesthetist and other members shall be from local

medical profession, non-government organization and Panchayati Raj Institution. At least, one member of this Committee shall be a woman. The tenure of the Committee is for two years. The non-government members shall have a maximum of two terms.

2. ***Experience***: For a RMP registered before the commencement of the Act, a minimum of three years professional experience in obstetrics and gynecology is compulsory. For those RMPs who are registered in State Medical Register and who hold a post-graduate degree or diploma in obstetrics and gynecology, the experience gained during training is sufficient.

   In case of a medical practitioner registered in a State Medical Register, six months of house surgency in obstetrics and gynecology is adequate. If he / she has assisted a RMP in performance of twenty five cases of medical termination of pregnancy of which at least, five have been performed independently in an approved hospital, he / she is permitted to conduct termination of pregnancy only upto 12 weeks of gestation.

3. ***Approval of Place:*** The State Government should be satisfied that termination of pregnancies shall be done under safe and hygienic conditions. For termination upto 12 weeks of pregnancy, the place should have a gynecology examination / labour table, resuscitation and sterilization equipment, facilities for transport and back up facilities for transport and essential drugs and parental fluid.

   In case of termination of pregnancy upto 20 weeks, an operation table, instruments for gynecological Surgery, anesthetic, resuscitation and sterilization equipment, and essential drugs and parental fluids for emergency use as notified by Government of India from time to time are mandatory requirements.

   The certificate of approval issued by the Committee shall be displaced at the place to be easily visible to persons visiting the place.

CHAPTER 14

Essential Commodities Act, 1955

It is an Act to control production, supply, distribution, and commerce of certain commodities. This Act was enacted in 1955 to protect the interests of consumers, with an objective of providing essential commodities that are in short supply at a reasonable price.

Essential Commodities

The essential commodities covered under this Act are:

1. Aluminum, iron, steel and copper
2. Cement
3. Coal and cinema carbon
4. Cotton and cotton textile
5. Dry batteries
6. Drugs
7. Fertilizers
8. Foodgrains, maize, pulses, edible oils, gur, salts, sugar, tea, wheat, wheat products and vegetable oil products
9. Fruit products and meat products
10. Insecticides
11. Jute
12. Petroleum, diesel oil, furnace oil, lubricating oil, petroleum gas and kerosene

13. Newsprint, paper, paper board and straw board
14. Power crushers and seeds
15. Pressure stoves, household electric appliances and electric lamps
16. Synthetic rubber, tyres and tubes
17. Any commodities which the Government notifies in the Official Gazette.

Control of Production, Supply and Distribution of Essential Commodities

For maintaining and/or increasing supplies of any essential commodity or for securing their uniform distribution and availability at fair prices, the Government provides for regulating or prohibiting production, supply, distribution and trade and commerce thereof. The Act provides for –

1. Regulating manufacture or production of essential commodities.
2. Regulating storage, transport, distribution, disposal, acquisition, use or consumption of any essential commodity by the licence holder.
3. Controlling the price of essential commodity at which it may be bought or sold. The drugs (Price Control) orders are made under this Act.
4. Prohibiting or increasing the sale of any essential commodities.
5. Regulating or prohibiting any class of commercial or financial transactions relating to foodstuffs or cotton textiles which may be detrimental to the public interest.
6. Maintaining accounts, records and registers and submitting them to the inspecting authorities by persons engaged in production and sale of essential commodities.
7. Collecting any information or statistics regarding activities related to essential commodities.
8. Bringing any waste or arable land under cultivation for growing or increasing the cultivation of food crops including, sugarcane on such lands.

Pricing of Essential Commodity

In accordance with provisions of this Act, any person holding in stock or engaged in the production or involved in the business of buying or selling of any essential commodity is required to sell the whole or a specified quantity to the Central or State Government or any authority of Government as and when asked for by a specified order. Such a person who sells any essential commodity in whole or a specified quantity to the Central or State Government or to an authorized agency shall be paid the price thereof as follows:

(i) Where the price is fixed by an agreement, the agreed price.

(ii) Where no such agreement exists, the price worked out with reference to the controlled price.

(iii) In absence of (i) and (ii), the average market rate prevailing in the locality during the period of three months preceding the date of notification.

Confiscation of Essential Commodity

1. A Government Officer, inspecting the premises and suspecting that contravention of rules and regulations is taking place, shall inform the matter to the Collector immediately. The Collector inspects that premises and satisfies himself that the contravention of the Act is taking place as reported by the Government officer. The Collector may order confiscation of –

   (i) the essential commodity or

   (ii) any package, covering or receptacle in which such essential commodity is found or

   (iii) any animal, vehicle, vessel or other conveyance used in carrying such essential commodity.

2. The confiscation of any essential commodity shall be made only after being served with a notice in writing, informing the person of the grounds on which it is proposed to confiscate.

   If the confiscated essential commodity is likely to undergo speedy and natural decay the Collector may order the same to be sold at the fixed controlled price, if any, fixed for such essential commodity under this Act or under any law for the time being in force, If no such price is fixed, the Collector may order the same to be sold by public auction, or in case of foodgrains, the same may be sold through fair price shops at the price fixed by the Central or State Government for the retail sale.

3. In such cases after deduction of Government expenses, the remaining sale amount of confiscated articles shall be paid to the owner or the person from whom it is seized, if –
   (i) no order of confiscation is ultimately passed by the Collector or
   (ii) where no order is passed on appeal or
   (iii) wherein a prosecution instituted for contravention of the order, the person concerned is ultimately acquitted.
4. Any aggrieved person by an order of confiscation, may appeal to the judicial authority appointed by the State Government within thirty days from the date of communication.

Penalties

1. For contravention of the provision of the Act, a person shall be punishable with imprisonment upto one year and also shall be liable to fine.
2. A person preventing collection of any information or statistics or failing to produce the records needed to be maintained under the Act to the concerned authorities, shall be punishable with atleast three months imprisonment which may extend to seven years and shall also be liable to a fine if he/she violates any order under the Act.

Chapter 15

Insecticides Act, 1968

Introduction

It is an Act to regulate the import, manufacture, sale and distributions, transport and use of insecticides with an objective of preventing risk to human beings or animals and for matters connected therewith. The Act extends to the whole of India.

The Central Government notifies dates for commencement of the Act and its provisions in different states.

The sprays of insecticides used for killing insects on the plants may leave the residue which may be poisonous and carcinogenic. The use of insecticides for household and agricultural purposes if not regulated properly, may pose a potential health hazard.

Definitions

1. **Insecticide** – Any substance listed in the Schedule to the Insecticides Act or any preparation containing any one or more of such substances.
2. **Misbranded Insecticide** –

   Any insecticide which is –

   (i) Not registered in prescribed manner.

   (ii) Not packed or labelled in accordance with the regulations.

   (iii) Sold under the name of another insecticide.

(iv) Imitation of genuine insecticide.

(v) Falsely labelled and/or deceptive in relation to its contents.

(vi) Possesses higher toxicity than the prescribed one.

(vii) Mixed or packed with other substances for altering quality or containing any unregistered substance.

Administrative Agencies

The following authorities/agencies are established under the Act.

(I) Central Insecticide Board

(II) Registration Committee

(III) Central Insecticides Laboratory

(IV) Insecticide Analyst

(V) Insecticide Inspector

(I) Central Insecticide Board (CIB)

It is constituted by the Central Government for advising Central and State Governments on all technical matters arising out of the administration of the Act, It consists of ex-officio and other nominated members.

[A] Ex-officio Members (17)

1. Director-General of Health Services (Chairman)
2. Drugs Controller General of India
3. Director General, Indian Council of Agricultural Research, (ICAR)
4. Director General, Indian Council of Medical Research, (ICMR)
5. Director General, Indian Standards Institution (ISI)
6. Director General of Shipping (Ministry of Transport and Shipping)
7. Plant Protection Adviser to the Government of India
8. Director, National Institute of Communicable Diseases
9. Director, Storage and Inspection, Ministry of Food, Agriculture, Community Development and Co-operation.
10. Director, Zoological Survey of India
11. Chief Adviser of Factories, Government of India
12. Secretary, Central Committee for Food Standards
13. Industrial Adviser (Chemicals), Directorate General of Technical Development

14. Joint-Director, Traffic (Ministry of Transport and Shipping)
15. Commissioner of Animal Husbandry
16. Joint Commissioner (Fisheries), Department of Agriculture, Government of India
17. Deputy Inspector General of Forests (Wild Life) Department of Agriculture

[B] Nominated Members (7)

1. One person representing Ministry of Petroleum, Government of India
2. One medical toxicologist
3. One pharmacologist
4. One person who is in charge of the department dealing with public health in a state
5. An expert in industrial health and occupational hazards
6. One person representing the Council of Scientific and Industrial Research (CSIR)
7. One ecologist

Functions of the Central Insecticide Board (CIB)

1. The CIB advises the Central Government on technical matters in relation to
   (i) The manufacture, transport, distribution, sale and storage of insecticides, keeping in view the safety to human beings or animals.
   (ii) The risk to human beings or animals involved in the use of insecticides, and the safety measures to prevent such a risk.
2. The shelf-life of insecticides.
3. The uses of, and classification of insecticides on the basis of their toxicity and suitability for aerial application.
4. The colouring matter to be mixed with concentrates of insecticides (particularly highly toxic insecticides).
5. The tolerance limits for pesticide-residues and on establishment of minimum intervals between the application of insecticides and harvesting.

6. Other functions supplemental, incidental or consequental to any of the above functions in accordance with the Act or Rules.

(II) Registration Committee

The Registration Committee constituted by the Central Government shall regulate its own procedure and the conduct of the business to be transacted by it. It consists of the following members.

(i) The Chairman and

(ii) Not more than five persons from the Central Insecticide Board including, the Drugs Controller, India and Plant Protection Adviser to the Government of India.

The Chairman is not from the members of Central Insecticide Board.

Functions of Registration Committee

1. Registration of insecticide after careful scrutiny of its formulae and verifying the claims made by the manufacturer or importer regarding its efficacy, and safety to human beings and animals.
2. Specification of the precaution to be taken to avoid poisoning through the use of handling of insecticides.
3. Other functions, supplemental, incidental or consequental to any of the functions assigned by the Act or Rules.

(III) Central Insecticides Laboratory

The Central Government by notification establishes a Central Insecticides Laboratory under the control of a Director.

The functions of Central Insecticide Laboratory shall also be carried out at any such institution as may be specified.

Functions of Central Insecticides Laboratory (CIL)

The CIL is responsi1ble for –

1. The analysis of the samples of insecticides sent by the officers authorised by the Central or State Government and submission of the certificate of analysis to the concerned officer or authority.
2. The analysis of samples of materials for insecticide residues under the provisions of the Act.

3. Determining the efficacy and toxicity of insecticides.
4. Carrying out such investigations as may be necessary for the purpose of ensuring the conditions of Registration of insecticides.
5. Undertaking other activities entrusted by Central Government or by a State Government with the permission of Central Government and after consultation with the Central Insecticide Board.

(IV) Insecticide Analysts

The Central Government or a State Government may appoint Insecticide Analysts for such areas and in respect of such insecticides or class of insecticides as may specified from time to time.

No person with financial interest in the import, manufacture or sale of any insecticide shall be appointed as Insecticide Analyst.

Qualifications for Insecticide Analyst

A graduate in agriculture or in science with chemistry as a special subject with adequate training in analysis of insecticides in a recognized laboratory.

Duties of Insecticide Analyst

1. To analyse or to test or to cause to be analysed samples of insecticides sent by the Insecticide Inspector under the provisions of this Act and to furnish the reports of such tests or analsys.
2. To forward reports from time to time to the State Government regarding the results of analytical work and investigation for publication.

(V) Insecticides Inspectors

The Central Government or a State Government may appoint Insecticide Inspectors for such areas as may be specified in the notification.

The person to be appointed as an Insecticide Inspector should not have any financial interest in the import, manufacture or sale of any insecticide.

Qualification of Insecticide Inspector

Graduate in agriculture, or graduate in science with chemistry as one of the subjects with adequate field experience.

Duties of Insecticide Inspector for Inspection of sale

(i) To inspect all establishments selling insecticides within the area of his/her jurisdiction.

(ii) To satisfy himself/herself by visiting establishment atleast thrice a year that the conditions of licence are complied with.

(iii) To procure and send the samples of insecticide for test and analysis.

(iv) To investigate any complaint sent to him in writing.

(v) To maintain records of all inspections made and actions taken by him/her.

(vi) To institute prosecutions in respect of breaches of the Act and the Rules thereunder.

(vii) To make such inquires and inspections as may be necessary to detect the sale and use of insecticides in contravention of the Act.

Duties of Inspectors Authorised to Inspect Manufacture of Insecticides

1. To inspect the premises licenced for the manufacture of insecticides within the area of his/her jurisdiction atleast twice a year.
2. To satisfy himself/herself that the conditions of the license and the provisions of the Act or Rules made thereunder are complied with.
3. To forward the detailed inspection report to the licensing officer.

Registration of Insecticides

For the import or manufacture of an insecticide, registration is mandatory. An application in prescribed format has to be made for registration to the Registration Committee appointed under the Act. If the Committee is satisfied, it allots a registration number and issues a certificate to that effect.

In case of refusal to register, an aggrieved person may appeal within thirty days to the Central Government whose decision shall be final and binding.

Grant of Licence

The licensing officers are appointed by State Government for specified areas to grant the licences to manufacture or to sell, stock or exhibit for sale or distribute any insecticide.

Any individual, who wishes to manufacture or sell or distribute or stock for sale any insecticide, has to apply to the licensing officer in prescribed form for grant of licence.

On receipt of such application, the licensing officer may grant a licence on prescribed form on such conditions prescribed and on payment of such fees as may be prescribed.

The licence granted is valid for the specified period and may be renewed from time to time on payment of such fees as may be prescribed.

Listed Insecticides

The insecticides listed in the Schedule are as follows.

- Aldrin
- Arsenates of calcium and copper
- Carbon tetrachloride
- Carboxin (DCMO)
- Copper Sulphate
- Cyanides of calcium and copper
- DDD (Dichloro Diphenyl Dichloroethane)
- DDT (Dichloro Diphenyl Trichloroethane)
- Endrin
- Malathion
- Methyl bromide
- Parathion
- Potassium Cyanide
- Pyrethrins
- Strychnine
- Tetraethyl pyrophosphate

Offences and Penalties

1. A person who –
   (i) Imports, manufacture, sells, stocks or exhibits for sale or distributes mis-branded insecticide or
   (ii) Imports or manufactures insecticide without a certificate of registration and
   (iii) Manufactures sells, stock or distributes without a licence or
   (iv) Obstructs an insecticide inspector in the performance of his/her duties, or
   (v) Causes an insecticide to be used by any worker in contravention of the provision of the Act is punishable with imprisonment upto two years or fine upto two thousand rupees or both for first conviction. For second and subsequent conviction, the punishment is imprisonment upto three years or fine or both.
2. A person using an insecticide in contravention of any provision of this Act or any Rules made thereunder is punishable with fine upto five hundred rupees.
3. Any person who contravens-
   (i) Any other provisions of this Act or Rules thereunder or
   (ii) Any condition of certificate or registration or licence, is punishable with imprisonment upto six months or with fine or with both for first offence For second and subsequent offence, punishment is imprisonment upto one year or fine or both.
4. If the offence is committed by the company, the incharge of the company who is responsible for the conduct of it's business is deemed to be guilty.

   However, if the person incharge proves that he/she has taken due care to avoid the offence, he/she, may be excused.

CHAPTER 16

The Consumer Protection Act, 1986

Introduction

The Consumer Protection Act, 1986 was enacted with the sole objective of providing better protection of the interests of consumers through establishment of Consumer Councils and other Authorities for the settlement of consumer disputes and connected matters. It came into force from 24th December, 1986. The Act extends to the whole of India except, the state of Jammu and Kashmir. The Rules were framed in 1987 and major amendment to the Act has taken place in 2002.

The Consumer Protection Act, 1986 is an important milestone in the history of socio-economic legislation in India. The Act has facilitated consolidation of the processes of consumer protection. It has opened a new chapter in consumer jurisprudence.

Hon. Supreme Court has directed that medical services are covered under the definition of "services" for the purpose of this Act. Therefore, rendering of consultation, diagnosis and treatment (both medical and surgical) are covered under the provisions of this Act. This is how the services of pharmacists are accountable in relation to consumer protection.

Definitions

1. Consumer

A person who –

(i) buys any goods for a consideration which has been paid or promised or partly paid and partly promised, or is under any system of deferred payment. It also includes any user of such goods other than the person who buys such goods for consideration paid or promised or partly paid or partly promised or is under any system of deferred payment when such use is made with the approval of such person. However, it does not include a person who procures such goods for resale or for any other commercial purpose

or

(ii) hires or utilizes any services for a consideration which has been paid or promised or partly paid and partly promised, or is under any system of deferred payment. It also includes a beneficiary of such services other than the person who hires or avails of the services for consideration paid or promised, or partly paid and partly promised, or is under any system of deferred payment, when such services are availed of with the approval of the first mentioned person.

2. Complainant

The following categories are covered –

(i) a consumer or

(ii) one or more consumers, where there are numerous consumers having the same interest or

(iii) the Central Government or any State Government who makes a complaint Or

(iv) any voluntary consumer association registered under the Companies Act, 1956 (1 of 1956), or under any other law for the time being in force.

3. Complaint

Any allegation in writing made by a complainant that –

(i) an unfair trade practice or a restrictive trade practice has been adopted by any trader.

(ii) the goods bought by him or agreed to be bought by him suffer from defect.

(iii) the deficiency in services hired or availed of or agreed to be hired or availed

(iv) excess price charged for the article.

(v) hazardous good being offered for use.

4. **Restrictive Trade Practice**

   It refers to a trade practice which requires a consumer to buy, hire or avail of any goods or, as the case may be, services as a condition precedent for buying, hiring, or availing of other goods or services.

5. **Defect**

   Any fault, imperfection or shortcoming in the quality, quantity, potency, purity or standard in accordance with law.

6. **Deficiency**

   Any fault, imperfection or shortcoming or inadequacy in the quality, nature and manner of performance in accordance with law.

Central Consumer Protection Council (CCPC)

The Central Government establishes the Central Consumer Protection Council for promotion and protection of the rights of the consumer. The State Government also establishes the State Council for the respective State to promote and protect the rights of the consumers.

In accordance with the provisions of the Act, the following Consumer Dispute Redressal Agencies have been set up:

(i) National Consumer Disputes Redressal Commission established by the Central Government.

(ii) The "State Commission" established by the State Government in the State and

(iii) The "District Forum" established by the State Government in each district.

The National Commission has the administrative control over all the State Commissions. The State Commission shall have administrative control over all the District Fora within its jurisdiction.

Where a trader or a person against whom a complaint is made or the complainant fails or omits to comply with any order made by the District Forum, the State Commission or the National Commission, such a trader or the complainant shall be punishable with imprisonment of one month to three years or with fine between ‘two to ten thousand.

For a frivolous or vexatious complaint made by the complainant to any of the Consumer Dispute Redressal Agencies, the complainant will be ordered to pay to the opposite party fine not exceeding ‘ten thousand.

There shall be no prosecution against any member of District Forum, State Commission or National Commission for executing any order, if such an order is taken out in good faith.

Chapter 17

Poisons Act, 1919

It is the Act enacted in 1919 to consolidate and amend the law regulating importation, possession and sale of poisons. It extends to whole of India. but is not applicable to the state of Jammu and Kashmir except, to the importation of any specified poison in accordance with the provisions of the Act.

Empowerment

The Central Government has been empowered to regulate the import of poisons in India, whereas the State Governments have been authorized to make rules regarding the possession and sale of poison within their respective territories.

Poison

Any substance specified as a poison in a rule made or notification issued under this Act is deemed to be a poison.

Import of Poisons

The Central Government may prohibit the import of any specified poison across any defined customs frontier into India. However, it is permitted only under authority of the licence, in accordance with the conditions of the licence. Such a licence is granted by the Central Government for a particular duration and purpose.

Possession for Sale and Resale of Poisons

The State Government makes rules to regulate the possession and sale of poisons, whether wholesale or retail, within whole or any part of the territories under its administration. The rules prepared for the purpose provide for –

(i) The grant of licences to possess specified poison for sale (wholesale or retail) and fixing of the fee to be paid.

(ii) The category of persons to whom the licences are to be granted.

(iii) The classes of persons to whom such poisons may be sold.

(iv) The maximum quantity of poison that could be sold.

(v) The maintenance of a sales register by the person who has granted licence.

(vi) The safe custody of such poisons and proper labelling of vessels, packages or coverings.

(vii) The inspection of poisons and the registers maintained for their sale by persons authorized by the State Government.

The State Government is also empowered to regulate the possession of any specified poison in any local area, where such poison is frequently used for mischief or murder or for poisoning the cattles.

Issue of Warrants

The District Magistrate, the Sub-divisional Magistrate or Commissioner of Police have the power to issue warrants for the search of any place in which he/she has reason to believe that any poison is possessed or sold in contravention of the Act or any rule thereunder. The search is undertaken by the authorized person in accordance with the provisions of the Code of Criminal Procedure relating to search warrants.

Penalties for Offences

The person who imports, possesses or sales any poison in contravention of the Act is punishable with imprisonment upto three months or fine upto '.500 or both on first conviction. The subsequent conviction attracts penalty of imprisonment upto six months or a fine upto ' One thousand or both.

Rules

The State Government may also make additional rules to carry out the purposes and objects of this Act, except for import of poisons. Such rules are published in the Official Gazette.

Savings

The provisions of the Act shall not apply to anything done in good faith by a medical or veterinary practitioner.

The State Government may declare that all or any of the provisions of this Act (except import) shall be deemed not to apply to any article poison or class of articles poisons of commerce specified in such order.

CHAPTER 18

Shops and Establishments Act

This Act is enacted to ensure a fair deal to the employees in shops and other commercial establishment by their employers. The Acts in different States of India differ in matters but, the main provision and objects of these Acts remain the same. The Bombay shops and Establishment Act was enacted in 1946 and it extends to the States of Maharashtra and Gujarat. The object of the Act is to regulate conditions of work force in different establishments such as shops, restaurants, hotels, theatres commercial establishments, etc. In relation to pharmacy profession, the provisions of this Act are applicable to the pharmacists in retail and wholesale trade.

1. Registration of Establishments

1. All shops and commercial establishments are required to register with the Inspector of shops within thirty days of its commencement sending the prescribed fee and following details in the prescribed from:
    - (i) The name and postal address of the establishment
    - (ii) The name of the employer or the manager, if any
    - (iii) The category of the establishment (either a shop or other commercial establishment)
    - (iv) Number of employees working in the establishment
    - (v) Any other information as may be prescribed

2. The registration certificate, issued by the Inspector, has to be prominently displayed in the establishment and should be renewed at prescribed intervals.
3. In case of any change in the name, address, managership of the establishment or its closure, the owner must communicate the same to the Inspector within fifteen days.

2. Hours of Work

(i) For employee who are more than 18 years, the working hours are not more than 9 hours per day and not more than 48 hours in a week.

(ii) During periods of stock checking or preparation of accounts, the workers may work upto 54 hours a week.

(iii) The employees should be allowed an interval of atleast half an hour for meals and rest.

(iv) The employer is required to maintained a record of the hours worked by employees and of the intervals allowed to them for rest or meals.

(v) All employees who work beyond their working hours are required to be paid over-time at twice the rate of their normal remuneration.

3. Opening and Closing Times of Shops and Establishments

(a) In summer (April-September) – opening not before 7 A.M and closing not after 10.P.M.

(b) In winter (October-March) - opening not before 8 A.M and closing not after 9.P.M.

4. Closing day

Each shop or establishment shall be closed for one day in a week which will be decided by the owner. However, the pharmacies and chemists and druggists shops may remain open on a closing day.

5. Cleanliness, Ventilation, Lighting, etc., of the Establishments

The establishment should be kept clean, well-ventilated with protection from fire or any hazard. Drinkable water facility should be provided to the workers. The Inspector of shops is empowered to inspect the premises for required facilities.

6. Wages

(i) Every employer is required to fix the date for payment to the employees.

(ii) The wages must be paid in cash and should be paid within seven days of expiry of the wages. On termination of an employee, the wages due to him/her should be paid within a day of termination of the employment.

(iii) Employers, imposing fines or making deduction for loss or damages are required to explain the basis for imposing fines or deductions, personally and also in writing to the employees. The employer should be provided with an opportunity to explain his/her stand.

(iv) All fines and deductions must be recorded in a register maintained for the purpose. The amount should be utilised in accordance with the direction of the Government.

(v) The employees may make entitled deductions approved by Government from the wages of the employees.

(vi) If the employee prefers any claim against his/her employer for the non-payment or delayed payment of the wages, such claim should be made to the authority specified in this behalf by the Government within six months of the dispute. The authority after hearing the applicant and the employer, may direct the latter to pay the applicant, the amount due to him/her together with any compensation, subject to a maximum of half the sum due.

(vii) If the claim is found to be a fictitious and malicious one, the applicant may be directed to pay a penalty.

(viii) The decisions of the Authority shall be final.

7. Employment and Dismissal of Workers

1. The employee is required to provide a letter of appointment to every employee.
2. The employer can not dismiss the services for three months without giving him atleast one month's notice in writing or *in lieu* thereof, one month's wages.
3. The notice for dismissal is not required, if an employee is dismissed for misconduct, provided he/she is given an opportunity to explain himself/herself.

4. Those employees who have been in service for more than three months are required to give their employers one month's notice or one month's salary, *in lieu* thereof, if they want to leave the job.

8. Leave for the Employees

The employee is entitled for –

(i) Privilege leave with full wages for not less than 15 days after 12 months of continuous employment. The privilege leave may be accumulated upto a maximum of 30 days.

(ii) Sickness or casual leave for not more than 12 days in a year with full wages. This leave can not be accumulated.

9. Inspection for Establishments

1. The Government may appoint a Chief Inspector and such other Inspectors as may be necessary for the purpose of the Act.
2. The Chief Inspector is empowered to enter at all reasonable times any shop or establishment, with a view to inspect the premises and, in accordance with the provisions, examine different registers and records that are maintained under the Act.

10. Offences and Penalties

The penalties are imposed for following offences -

(i) For false entries and omissions of entries, a person is liable to imprisonment upto three months or a fine between '.50 and '.250 or both.

(ii) Any person causing obstruction to Inspectors in discharging his/ her duties or prevents an employee in the establishment from appearing before the Inspector, shall be liable to fine of '.50 to '.250.

(iii) Anyone who fails to maintain any record or display notice required to keep or displayed under the Act, shall be liable to a fine of '.50 for every day on which the contravention occurs.

(iv) Any person, who contravenes any other provision of the Act shall on conviction, be liable to punishment with a fine between '.25 and '.250.

CHAPTER 19

Prevention of Cruelty to Animals Act, 1960

Objective

This Act was enacted in 1960 to prevent the infliction of unnecessary pain or suffering on animals, as well as, to prevent cruelty to animals. The Act extends to the whole of India except, the state of Jammu and Kashmir. There are severe restrictions on use of animals for laboratory experiments. For establishing the therapeutic efficacy and safety of drugs for public use, experimentation with animals is sometimes unavoidable, but causing them unnecessary pain or suffering is both unethical and inhuman. This Act provides for preventing unnecessary pain and cruelty to animals.

All species of animals (except human being), as well as, all species of birds are covered for the purpose of Act. Cruelty means inflicting of unnecessary pain or suffering to animals.

The provisions of pharmacy Act related issues covering experimentation on animals are included in this Act.

Experimentation on Animals

1. Experimentation on animals is a tool for the advancement of pharmaceutical medical, and biological sciences since, animals are much similar to human system.
2. Animals are used for the production of sera, vaccines, hormones, enzymes and certain other biological.

All these processes require animals to be incised, dissected and sometimes, killed. From an aesthetic sense, all these experiments on animals should be carried out in such a manner that the animals do not suffer any avoidable pains and injuries. The Committee to look after the various aspects of experiments on animals and to supervise and control their use for experimentation thereby, saving them from any avoidable pain or injury, is constituted under the provisions of is Act.

The Committee comprises of the following members:

1. Nominated Members

(i) Two members each of the Indian Council of Medical Research (ICMR), Indian Council of Agricultural Research (ICAR), and Council of Scientific and Industrial Research (CSIR) nominated by the Central Government.

(ii) Two members representing universities granting degrees in medical and veterinary sciences nominated by the Central Government.

(iii) Five persons actively engaged in the promotion of animal welfare nominated by the Central Government.

2. Elected Members

One member each elected by the Lok Sabha and the Rajya Sabha.

The Committee is mandated to take all appropriate measures in order to ensure that experimental animals are not subjected to unnecessary pain or suffering before, during or after the experiments.

The Committee is authorized to make the rules under the Act which provide for the following.

(i) For experiments performed in an Institution, the Head of the Institution shall be responsible for ensuring compliance with the provisions of the Act. The experiments are required to be carried out only by qualified individuals. They shall be held responsible for avoidance of cruelty to the animals.

(ii) The use of large animals should be avoided where small animals such as, frogs, rats, rabbits, etc. can be used for an experiment.

(iii) Experiments on animals should not be performed for the sake of acquiring manual skill.

(iv) The models, films, charts, books, etc. should substitute animal experimentation, wherever possible.

(v) Animals intended to be used for experiments should be properly looked after before and after the experiments and records of experiments performed should be maintained.

(vi) The experiments should, as far as possible, be performed while the animals are under the influence of an anesthetic. If the animals are so injured during the experimentation that their recovery would involve serious suffering, they should be destroyed while still insensible.

Inspection

The Committee is empowered to authorise any of its officers or any other person to inspect any institution or place where experiments on animals are being carried out and to inspect the records required to be maintained under the provisions Act.

Prohibition of Experimentations

If the Committee is of the view that the person or institution is not carrying out experiments on animals in accordance with the prescribed rules, it is empowered to prohibit any other person or institution from carrying out experiments for a definite period of time or indefinitely. In such cases, it may also impose any special conditions for the performance of experiments.

Penalties

Any contravention of any order of the Committee or breach of any conditions imposed by the Committee is punishable with a fine extending upto '.200.

CHAPTER 20

The Prevention of Food Adulteration Act, 1954 and Rules, 1955

Food is an essential commodity required for the survival of all living organisms, including the human beings. Such an important commodity has to be properly protected by law in order to ensure its availability to consumers in hygienic conditions. The prevention of adulteration in foods and their products is the responsibility of the Government and ethical duty of the society. This Act is aimed at preventing import, manufacture, sale or distribution of adulterated and misbranded food and also to prevent all types of food adulterations. It is extended to whole of India. In many States, it is Food and Drug Administration (FDA) looking after the implementations of the provisions of this Act.

Definitions

Food

It means :

1. Any article used as food or drink for human consumption other than drugs and water and includes:
   (i) Any article which ordinarily enters into or is used in composition or preparation of human food.
   (ii) Any flavouring matter or condiment used in food preparation.
   (iii) Any other article which the Central Government may, having regard to its use, nature of the substance or quality, declare by notification in Official Gazette as food for purpose of this Act.

Adulterated Food

It means :

(i) The food that is not of nature, quality or substance demanded by the purchaser.

(ii) The food that contains a substance, which effects injuriously the nature or quality of the food.

(iii) Inferior or cheaper substance in place of authentic food affecting the quality of food.

(iv) A food article which is wholely or in part abstracted affecting the quality of food.

(v) The food article obtained from diseased animal.

(vi) The food article prepared, packed or kept under insanitary conditions.

(vii) The food article which consist wholely or in part filthy, putrid, rotten, decomposed or diseased animal or vegetable substance or is insect-infested and unfit for human consumption.

(viii) The food containing any coloring matter other than prescribed in the Act or if the amount of this prescribed coloring matter is not within prescribing limits.

(ix) A food article which contains prohibited preservative or permitted preservative in excess.

(x) A food article which falls below the prescribed standards.

(xi) A food article which contains any poisonous or other ingredient rendering injurious effects to human being.

Misbranded Food

It means :

(i) An imitation or a substitute resembling authentic food.

(ii) An article which is falsely stated to be the product of any place or country.

(iii) An article if sold by name which belongs to other article.

(iv) An article wherein the damage is concealed by coating, polishing or other operation.

(v) The article with false claims on label of the article.

(vi) An article with false labeling in any manner i.e., in terms of content, composition or even the owner.

(vii) Any artificial flavoring, coloring or chemical preservation of an article without declaration on the label.

Food (Health) Authority

The Director of Health and Medical Services or Chief Officer Incharge of Health Administration in the State or any officer empowered by Central or State Government shall be the Food (Health) Authority.

Local Area

Any area, urban or rural, identified as Local Area by notification in Official Gazette declared by Central or State Government.

Local Authority

For Local Areas-Municipal Board or Corporation or a Cantonment Authority or a notified area committee by Central or State Government.

Central Committee for Food Standards

It is constituted by the Central Government to advise Central or State Governments on all matters arising out of administration of this Act and also for carrying out other functions. The Director General of Medical and Health Services, Government of India is the Chairman of this Committee. The total number of members of the Committee is not indicated. The Committee can frame by-laws, constitute committees and meet as and when required.

Central Food Laboratories

Central Government has established four well-equipped laboratories with experienced personnel in food analysis at Kolkata, Gaziabad, Mysore and Pune. These laboratories are required to undertake: (i) Analysis of samples sent by officers of Central Government; (ii) Fixation of standards and quality control parameters for food articles; and (iii) Collaboration work with State laboratories for analysis and standardization. Many of the State Governments have their own laboratories for food analysis manned by public analysts who work in collaboration with the Central Laboratories.

The Central Government prescribes procedure, fees to be paid for analysis, proforma to be used, etc.

General Provisions

Any food article, which is adulterated, misbranded or prepared in contravention of the provisions of the Act, is prevented from import.

Manufacturing Prohibition

Any adulterated and misbranded food manufactured in contravention with provisions of the Act and manufactured when there is temporary prohibition is prevented from manufacturing. For certain categories of food articles, especially the stored foods, licence is required for manufacturing.

Analysis of Food

Public Analyst

The person recruited for this post should not have interest in manufacturing, import, sale or distribution of food. Central Government or State Government may appoint **public analyst** for local area.

Different analysts for different articles can also be appointed.

The person with following qualification is eligible for the post of public analyst.

1. A person who holds degree in science with chemistry/food technology/food and drugs/biochemistry or equivalent examination and not less than five years practical experience after graduation in analysis of food articles or holds masters degree in chemistry/food technology/biochemistry/food and drugs/microbiology or Associateship of Institution of Chemists (analysis of food) or any other equivalent qualification with a minimum of three years practical experience after postgraduation in analysis of food.
2. The person should have been declared qualified for appointment by the Board appointed by Government.

Food Inspector

Food Inspectors can be appointed by the Central Government or State Government for Local Areas identified. A person to be appointed as Food Inspector should have no interest in manufacturing, sale or distribution of food articles.

The person with following qualification is eligible for the post of Food Inspector.

1. A medical officer incharge of Health Administration of Local Area or
2. A graduate in medicine with a minimum of one month training in inspection of food sampling work in an institution approved by Central Government or State Government or
3. A graduate in science with chemistry/agriculture/pharmacy/ veterinary science/food technology/dairy technology/public health/diploma holder in food or dairy technology or equivalent qualification and a minimum of three months satisfactory practical training in inspection of food and training approved by Central Government or State Government.

Duties of Public Analysts

On receipt of sample from Food Inspector, the public analyst confirms the authenticity of the sample sent by him comparing the seal on it. He keeps the sample in safe custody before, during and after analysis and preserves it for submission into court, as and when required. He carries out the analysis of food samples sent by inspectors using approved methods, records the findings of analysis in specified manner and informs the Food Inspector about the analysis. The public analyst should also be in touch with the Central Food Laboratories for the purpose of consultation and dissemination of knowledge.

Duties and Powers of Food Inspectors

1. To inspect establishment for licence for manufacture, storage or sale of food article and satisfy that the conditions and provisions of the Act are complied with
2. To make enquiries on receipt of complaints about the quality of food and conduct inspection accordingly

3. To procure the sample of food and send to public analyst as and when required
4. To investigate into matters of specific complaints given in writing
5. To maintain the records of inspection properly and inform higher authorities about inspection .

The food inspector can prohibit sale of food article in public interest for a specific period, stop the vehicles carrying adulterated or misbranded food, seize the samples of food articles prepared in contravention of the provisions of the Act, enter the premises within the reasonable time of business, and seize the sample or break open any package containing adulterated or misbranded food. For seizing the sample, a notice in writing has to be given to the owner. Three parts of the seized material are prepared in presence of witnesses and sealed. One part is sent to public analyst and two parts to local authorities.

Permitted Colors in Food

Permitted colors are Chlorophyll, caramel, β-carotene, annatto, coal-tar dyes, saffron and curcumin. The colour content should not be more than 0.2 mg per kg of food material.

Antioxidants Permitted in Food

Tocopherol, ascorbic acid, lecithin.

Preservatives Permitted in Food

Class I : Common salt, sugar, dextrose, glucose, syrup, vinegar, honey and edible vegetable oil.

Class II : Benzoic acid and salts, sulphurous acids and salts, nitrates and nitrites of sodium and potasium. Sorbic acid and salts, methyl diacetate, sodium, potassium and calcium salts of lactic acid. Not more than one class II preservative is permitted.

No nitrate or nitrite preservative to be added to infant food.

The Food Inspector functioning beyond the scope of this Act is liabable for penalty of rupees five hundred.

Chapter 21

National Pharmaceutical Pricing Authority (NPPA)

It was first established in 1997 by the Government of India to fix or revise the prices of controlled bulk drugs and formulations in the country. It was to ensure the objectivity and transparency in drug pricing matters and for making available the controlled medicines at the costs determined by NPPA.

Composition: The officers of NPPA are the Chairman, Member-Secretary, Directors for various operations, Advisors, Deputy Directors, Under Secretary for Administration and Section officers.

Functions of NPPA

1. Implementation and enforcement of the provisions of the Drugs (Prices Control) Order.
2. Monitoring the availability of drugs and taking necessary actions for overcoming shortages, if any.
3. Collection and maintenance of data on production, import and export, market share of companies and their profits, etc for both bulk drugs and formulations.
4. Undertaking and/or sponsoring studies in relation to pricing of drugs and pharmaceuticals.
5. Dealing with all legal matters arising out of the decisions of NPPA.

6. Recruiting required number of officers and other staff members in accordance with the procedure laid down by the Central Government.
7. Rendering advice to the Central Government on changes or revisions in the drug policy.
8. Assisting the Central Government in parliamentary matters related to drug pricing.

Price Fixation

(i) **NPPA** is charged with the responsibility of price fixation/revision of bulk drugs.It identifies bulk drug for the purpose of fixation/ revision, collects the data based on information provided and actual plant visits, prepares actual cost statements and technical parameters, prepares estimated cost, calculates fair price of bulk drug and also determines maximum sale price of the bulk drug which is notified in Official Gazette.

(ii) **NPPA** also determines prices of formulations containing scheduled bulk drugs either based on applications of the manufacturers or on *suo motu* basis. The prices are fixed based on examination of technical parameters; study of prices of scheduled bulk drugs; excipient claims and examination of process loss (P.L.) conversion cost (C.C.), packaging charges (P.C.) and packaging material cost (P.M.). The P.L., C.C. and P.C. are considered as per the norms notified in the Gazette vide S.O.578(E) dated 13.7.1999. The PM cost is permitted as per the actual claim supported by invoices. Maximum allowable post-manufacturing expenses (MAPE) is given at 100% on the ex-factory cost for indigenous formulation, while it is 50% of the landed cost for imported formulation.

The retail prices is worked out using the formula –

$$\text{R.P.} = [\text{M.C} + \text{C.C.} + \text{P.M.} + \text{P.C.}] \times \left[1 + \frac{\text{MAPE}}{100}\right] + \text{E.D.}$$

Where, R.P.,M.C. and E.D. represent retail price, material cost and excise duty respectively and other symbols as mentioned earlier.

The statutory taxes for bulk drugs used in formulation are considered at the actuals and net of MODVAT. Allowances upto 8% on notified price of scheduled bulk drug is considered.

If the manufacturer does not apply for revision of formulation prices within 30 days of price reduction of bulk drug, *suo motu* revision is undertaken.

The ceiling prices (fixed or revised) are notified in the Gazette of India under para 9 of DPCO, 1995.

CHAPTER 22

National Blood Policy

Salient Features

In compliance with the directives of Hon. Supreme Court of India in 1996, modification and change in the blood transfusion service was necessitated. The National AIDS Control Organisation (NACO), Ministry of Health and Family Welfare, Government of India, has come out with National Blood Policy with an aim to ensure easy access and adequate supply of safe and quality blood, as well as blood components collected or procured from a voluntary donor in well protected and fully equipped premises under optimum conditions. The Policy also ensures blood transfusion under supervision of trained personnel for all citizens who need it through comprehensive, efficient and perfect quality management approach.

Objectives of the Policy

The following objectives are aimed at:

1. The firm commitment of the Government for providing adequate quantity of blood and its components or products. For achieving this target, the National Blood Transfusion Programme is developed. The **National Blood Transfusion Council** (NBTC) is constituted which shall co-ordinate with **State Blood Transfusion Councils** (SBTC) develop guidelines to define NGO-run blood centers and ensure involvement of other ministries and other health programmes.

   It shall be the responsibility of Drugs Controller General to look after enforcement of standards of blood and blood products. The SBTC shall organize the blood transfusion service through the network of Regional Blood Centers, Satellite Centers and other Government, Indian Red Cross and NGO-run Blood Centers.

In order to meet special requirements of Armed Forces in remote areas, necessary amendments shall be effected in the Drugs and Cosmetics Act and Rules.

2. In order to ensure proper development and reorganization of Blood Transfusion Service, Blood Transfusion Councils at Center and in States and Union Territories shall be financially strengthened by pooling resources from various existing programmes. International funding, cost recovery of blood/blood components and raising of funds for blood transfusion service shall be attempted.
3. The latest technology shall be made available for operating the blood transfusion service. The updating and effective enforcement of provisions of Drugs and Cosmetics Act and Rules, prescribing minimum standards for testing, processing and storage of blood/ blood products, establishment of a Vigilance Cell, introduction of Quality System Scheme in all blood centers, and appointment of Quality Assurance Manager for each Regional Blood Center are the steps to be initiated.
4. The activities for development of bio-safety guidelines and Generic Standard Operating Procedures and introduction of External Quality Assessment Scheme shall be undertaken. Each Blood Center shall be asked to develop its own Standard Operating Procedures on various aspects of blood banking.
5. Regular workshops and proficiency testing exercises shall be conducted for in-service persons.
6. Extensive awareness programmes for blood banking services shall be launched which include donor motivation for ensuring adequate availability of safe blood. The emphasis shall be on recruitment and retention of voluntary, non-remunerated blood donors, enrolment of safe donors, regular voluntary donors instead of, replacement donors and involvement of youth in blood donation. A counsellor shall be appointed for each Blood Center and updated donors directory shall be maintained.
7. Appropriate clinical use of blood and blood products shall be encouraged. Blood and blood products shall be transfused only when necessary. In addition, effective and efficient clinical use of

blood shall be promoted and National Guidelines on "Clinical Use of Blood" shall be made available. Whenever possible, the use of plasma expanders shall be promoted to minimize the use of blood. Guidelines for management of blood supply during natural and man-made disasters shall be made available.

8. Strengthening of trained manpower through human resource development shall be achieved by adopting focused strategy. Medical colleges shall be encouraged to start P.G. programme. (M.D. in Transfusion Medicine). A separate department of Transfusion Medicine shall be established in medical colleges. Technical training in transfusion medicine for nurses, pharmacists, technicians, donor organizers, donor recruitment officers and others shall be emphasized. In-service training programmes for drug inspectors and all categories of personnel working in blood centers shall be organized.
9. The research and development in the field of transfusion medicine and related technology shall be encouraged by providing research grants from Corpus fund for multi-centric research initiatives. Development of computer-based information and management systems shall be on priority agenda.
10. Adequate legislative and educational steps shall be taken to eliminate profiteering in blood banks. Fresh licences to stand-alone blood banks in private sector shall not be granted. The stringent punishment shall be accorded by amending existing provisions of Drugs and Cosmetics Rules for unauthorized/irregular practices in blood banking system. Provisions for affiliation with a licensed Blood Bank for procurement of blood for their patients shall be incorporated.

CHAPTER 23

Pharmaceutical Policy – 2002

Salient Features

On account of liberalization of Indian economy and in the context of globalization of trade in drugs and pharmaceuticals, the Government of India, Ministry of Chemicals and Fertilizers, Department of Chemicals, came out with Pharmaceutical Policy – 2002 to meet the challenges of Post-GATT era. It was necessary to comply with the requirements under the WTO Agreements for effective implementation of policy of product-oriented patents under new patent regime w.e.f. 1-1-2005. The need for new initiatives beyond enumerated in Drug Policy of 1986 was felt for making Indian pharmaceutical industry more internationally competitive. The basic objectives of Drug Policy of 1986 modified in 1994 still remain largely valid.

Objectives

The main objectives of the Pharmaceutical Policy-2002 are :

1. Ensuring abundant availability of good quality of drugs and pharmaceuticals at reasonable prices within the country.
2. Strengthening the capabilities for cost-effective production of quality drugs.
3. Creating an encouraging environment and a good network for introduction of new drugs and technologies by ensuring new investments in Indian pharmaceutical industry.
4. Encouraging export of drugs/drug formulations by reducing barriers to the trade in pharmaceutical sector.
5. Promoting rational use of pharmaceuticals.

6. Strengthening the system of quality control of drugs and pharmaceuticals.
7. Encouraging R & D ventures in pharmaceutical sector, especially with focus on diseases endemic or relevant to India.

The Pharmaceutical Research and Development Committee (PRDC) was constituted in 1999 under the Chairmanship of Director General of CSIR to make recommendations for strengthening research and development in pharmaceutical industry and to identify areas where Government support is required for domestic R & D efforts. The PRDC suggested following conditions (Gold Standards).

1. Investment of five percent turnover of industrial production in R & D.
2. Employment of at least 100 research scientists in R & D in India.
3. Investment of at least ₹ 10 crore per annum in innovative research in India, including new drug development, new delivery systems, etc.
4. A minimum of ten patents for research done in India.
5. Facilities for own and operate manufacturing in India.
6. Establishment of a Drug Development Promotion Foundation (DDPF) and Pharmaceutical Research and Development Support Foundation (PRDSF).

A Committee called the Drugs Price Control Review Committee (DPCRC) under the Chairmanship of the Secretary, Department of Chemicals and Petrochemicals was set up in 1999. The recommendations in the report of this Committee have been examined while formulating "Pharmaceutical Policy-2002".

Keeping in view the interest of the weaker sections of the society, Government reserves the right of intervening in cases where prices behave abnormally.

The Central Government has taken following important decisions:

1. Industrial licensing for all bulk drugs cleared by Drugs Controller General of India shall be abolished except for bulk drugs produced by recombinant DNA technology and bulk drugs requiring *invivo* use of nucleic acids.
2. Foreign investment up to 100 percent will be permitted, subject to stipulations laid down in Industrial Policy.

3. Imports of drugs and pharmaceuticals will be as per the EXIM policy in force.
4. Liberalized approval policy for foreign technology agreements.
5. Establishment of Pharmaceutical Research and Development Support Fund (PRDSF) under the control of DST.
6. Constitution of Drug Development Promotion Board to administer utilization of PRDSF.
7. Provision of fiscal incentives to promote indigenous R & D efforts.
8. Two hundred and seventy-nine items appearing in National Essential Drug List, 1996 of Ministry of Health and Family Welfare and other 173 items considered important by the same ministry in various health programmes, emergency care, etc., shall be for price regulation. The ORG-MARG data shall be the basis for determining the span of price control as suggested by Drugs Price Control Review Committee (DPCRC)
9. Bulk drugs shall be under price regulation if,
   (i) Moving Annual Total (MAT) value for the drug is more than ₹ 25 crore and percentage share is 50 percent or more.
   (ii) The total MAT value for a particular drug is less than ₹ 25 crore but more than ₹ 10 crore and percentage share is 90 percent or more.
10. Maximum Allowable Post-manufacturing Expenses (MAPE) will be 100 percent for indigenously manufactured formulations.
11. For imported formulations, the margin covering selling and distribution expenses, including interest and importers' profit, shall not be more than 50 percent of the landed cost.
12. The time frame of two months from the date of the receipt of complete prescribed information for granting price approvals shall be fixed.
13. Ceiling prices of any formulation may be fixed which will be obligatory for all to follow.
14. A new drug developed through indigenous R & D efforts would be exempted from price control for fifteen years.
15. A formulation involving a new delivery system developed through indigenous R & D and patented shall be exempted from price control till the expiry of the patent.

16. The NPPA (National Pharmaceutical Pricing Authority) will be authorized to exempt such formulation from price control, if it's cost to consumer-patient does not exceed ₹ 2/- per day.
17. The NPPA will be revamped and reoriented to monitor prices of decontrolled drugs and formulations and oversee implementation of DPCO.
18. It shall be mandatory for manufacturer to furnish all information called by NPPA.
19. The Government's endeavor will be to upgrade the standards of pharmacy education and R & D through NIPER (National Institute of Pharmaceutical Education Research), Mohali, Punjab.

CHAPTER 24

The Drugs (Price Control) Order (DPCO), 1995

(Under Section 3 of the Essential Commodities Act, 1955 as published in the Gazette of India C, Extraordinary Part II Section 3, Sub-Section (ii) dated 6.1.1991 with amendments.)

The drug is categorized as essential commodity under Section 3 of the Essential Commodities Act, 1955 and certain categories of the drugs are kept under price control.

Based on New Drug Policy and Pharmaceutical Policy 2002 framed to meet the challenges of Post-GATT era, the DPCO was modified in order to facilitate the process of globalization of pharmaceutical trade and ensure production of quality medicines. The new patent regime in post-WTO era has thrown new challenge in the field of drugs and pharmaceuticals.

It is to make Indian pharmaceutical companies globally competitive, amendments were made to DPCO, from time to time, and several concessions were offered for promotion of indigenous R & D efforts. DPCO is a Government Order notified by Ministry of Chemicals and Fertilizers, Department of Chemicals, dated 6.1.1995 for regulating the price structure of certain categories of drugs and their formulations including life-saving drugs.

1. Important Definitions

(i) ***Bulk Drug :*** It means any pharmaceutical, chemical, biological or plant product or medicinal gas conforming to pharmacopoeial or other standard accepted under DCA, 1940 which is used in any formulations.

(ii) ***Formulation :*** It means a medicine processed out of or containing one or more of bulk drugs with or without the use of pharmaceutical aids for internal or external use for or in the diagnosis, treatment, mitigation, or prevention of disease in human beings or animals, but shall not include any bonafide Ayurvedic, Siddha, Unani, Tibbi and Homeopathic Systems of Medicine. It also does not include any substance to which provisions of DCA, 1940 do not apply.

(iii) ***Ceiling Price :*** This is the price fixed by Government for Scheduled formulations in accordance with the provisions of the Order.

(iv) ***Drug:*** It includes

(a) All medicines for internal or external use of human beings or animals and all substances useful in diagnosis, treatment, mitigation or prevention of any disease or disorder in human beings or animals, including insect-repellent preparations,

(b) Any substance capable of affecting any function of human or animal body or used for destruction of insect or vermin, which cause disease in human beings or animals, and

(c) Bulk drugs and their formulations.

(v) ***Manufacture :***

In relation to any drug under DPCO includes any process or part of a process for making, altering, finishing, packing, labelling, breaking up or otherwise, treating, or adapting any drug with a view to its sale and distribution, but doesn't include the compounding or dispensing or packing of any drug in the course of retail business.

(vi) ***Scheduled Bulk Drug :*** It means the bulk drug specified in First schedule to the DPCO.

(vii) ***Non-Scheduled Bulk Drug :*** It means a bulk drug not specified in the First Schedule to DPCO.

(viii) ***Pre-tax Return:*** It means profits before payment of income-tax and sur-tax and includes such other expenses as do not form part of the cost of the formulation.

(ix) ***Sale Turnover :*** It means the product of units of formulation sold by manufacturer or an importer, as the case may be, in one accounting year, multiplied by retail price inclusive of sales tax, if any.

(x) ***Retailer :*** It means a dealer carrying on the retail business of sale of drugs and cosmetics.

(xi) ***Wholesaler :*** It means a dealer or his agent or a stockist appointed by a manufacturer or an importer for the sale of his drugs to a retailer, hospital, dispensary, medical educational or research institution purchasing bulk quantities of drugs.

The First Schedule

It consists of the list of bulk drugs, around 74 used in the treatment of different ailments. The drugs include – sulphamethoxazole, penicillins, tetracycline, analgin, doxycycline, ciprofloxacin, rifampicin, cefotaxine, streptomycin, ranitidine, dexamethasone, ephedrine, vitamins A, B1, B2, C & E, betamethasone, carbamazepine, metronidazole, chloroquine, theophylline, insulin, erythromycin, levo-dopa, tolnaftate, oxytetracycline, nalidixic acid, prednisolone, griseofulvin, cephazolin, gentamycin, methyl dopa, aspirin, halogenated hydroxyquinoline, dextropropoxyphene, trimethoprim, cloxacillin, pentazocine, sulphadimidine, captopril, salbutamol, naproxen, pyrental, sulphadoxine, famotidine, ibuprofen, norfloxacin, frusemide, cafadroxyl, pheniramine maleate, panthenols and panthonates, chloroxylenols, furazolidone, becampicillin, pyrithioxine, lincomycin, chlorpropamide, sulphadiazine, framycetin, mebhydroline, chlorpromazine, verapamil, methendienone, phenylbutazone, glipizide, lynestranol, spironolactone, salazosulphapyrine, pentoxyfylline, diosmine, amodiaquin, trimipramine, and sulphamoxole.

The Second Schedule

It consists of different forms to be submitted such as (i) Form of information/application for fixation or revision of prices of scheduled bulk drugs (in duplicate), (ii) Form of information in respect of price of non-scheduled bulk drugs (in duplicate), (iii) Form of application for approval or revision of price of Scheduled formulations (seven copies), (iv) Form of application for approval or revision of price of Scheduled formulations imported in finished form (seven copies), (v) Form of Price List, and (vi) Yearly information on turnover and allocation of sales and expenses.

The Third Schedule

It specifies maximum pre-tax return on sales turnover of manufacturers or importers of different formulations. There are three categories of units for this purpose.

(i) Category A : Units with turnover of more than ₹ 6 crore per annum.

(ii) Category B : Units with turnover between ₹ 1 crore to ₹ 6 crore per annum.

(iii) Category C : Units with turnover of less than ₹ 1 crore per annum.

The Central Government has the power to fix maximum sale price of indigenously manufactured bulk drugs specified in First schedule. While fixing the price of a bulk drug, the Government takes into consideration a post-tax return of fourteen percent or a return of 22 percent on capital employed or in respect of a new plant an internal rate of return of twelve percent. Where the production is from basic stage, the Government shall take into consideration post-tax return of eighteen percent on net worth or a return of 26 percent on capital employed.

Time Limits

Fifteen days time for new bulk drug (First Schedule) production by the manufacturer, as well as for implementation of policy matters was announced by the Government. An application for revision of maximum sale price of bulk drug in **Form I** submitted by the manufacturer, shall be responded by the Government after careful study within four months from the date of receipt of complete information.

Every manufacturer shall furnish within 30 days of commencement of this order a list of all Scheduled drugs produced by him to the Government in **Form I** and that of non-scheduled drugs in **Form II**.

Both for bulk drugs and formulations, the manufacturer is required to submit the price list every year within one month of introduction of annual Finance Bill to the Government.

The Government may direct any manufacturer of any bulk drug to sell such drug to other manufacturers of formulations in order to ensure adequate and equitable distribution. The Central Government has to look into the requirements for captive consumption of such manufacturer and also to think of plant growth of pharmaceutical industry.

Calculation of retail price of formulation

The retail price of a formulation is calculated by the government using following formula.

$$R.P. = (M.C + C.C. + P.M. + P.C.) \times (1 + MAPE/100) + E.D,$$

where,

R. P. = retail price

M.C. = Material cost and includes cost of drugs and other pharmaceutical aids used including averages and process loss specified as a norm from time to time.

C.C. = Conversion cost worked out in accordance with established procedures of costing and fixed as a norm every year.

P.M. = Cost of packing material of formulation, including process loss and shall be fixed as a norm every year.

P.C. = Packing charges worked out in accordance with established procedures of costing and fixed as a norm every year.

MAPE = (Maximum Allowable Post-manufacturing Expenses) – All costs incurred by manufacturer from ex-factory to retailing stage and includes margin for manufacturer and trade margin and it shall not be more than one hundred per cent for indigenously manufactured Scheduled formulation.

E.D. = Excise duty to be paid.

In case of imported formulation the landed cost shall form the basis for fixing its price along with margin to cover selling and distribution expenses and importer's profit which shall not exceed fifty per cent of landed cost.

Power to Fix Retail Price

The Government may, from time to time, fix the retail price of a Scheduled formulation in accordance with the formula stated above. The price of formulation once fixed shall not be increased by the manufacturer except with the prior approval of the Government.

The application made by the manufacturer for revision of retail price in **Form III** or **Form IV**, as the case may bc, with valid justification shall be attended to by the Government within two months of submission of complete information. After scrutiny and study, if the claim is found justified, the Government may pass necessary order for revision of retail price.

No manufacturer shall market a new pack or sell or dispose of any imported Scheduled formulation without permission of the Government.

Power to Fix Ceiling Price

The Government may, from time to time, fix the ceiling price of a Scheduled formulation in accordance with the formula laid down, keeping in view the cost or efficiency or both of major manufacturers of such formulations.

The Government, after calling for necessary information, on its own or on application of manufacturer, after careful study, may fix revised ceiling price of a Scheduled formulation.

The Government after obtaining necessary information from manufacturer or importer fix or revise the retail price of one or more formulations marketed by such manufacturer or importer including, a non-Scheduled formulation, provided pre-tax return on sales turnover does not exceed maximum pre-tax return specified in the Third Schedule.

The Government reserves the right of fixing or revising retail price of any formulation including, non-Scheduled formulation and also fix or revise the prices of bulk drug and its formulation in accordance with the provisions of the Order.

Recovery of Dues

The Government may by notice direct the manufacturer-importer or distributor to deposit amount accrued under DPCO 1979 on or before commencement of this order into Drugs Prices Equalization Account within specified time. The amount shall be utilized for promoting pharmaceutical education and research, paying the shortfall between retention price and common selling price to the manufacturer, importer or distributor; or for meeting expenses incurred by the Government in discharging different functions related to the Order.

The manufacturer, importer or distributor has to display prices of non-Scheduled formulations and their price list. The manufacturer or distributor shall not refuse sale of drug to dealer and dealer, in turn, to a customer without sufficient reasons.

A manufacturer, distributor or wholesaler shall sell a formulation to a retailer unless otherwise permitted at a price equal to the retail price as specified by an order (excluding exise duty, if any) minus sixteen percent thereof for Scheduled drugs.

The Government reserves the right of fixing by general or specific order in public interest the price of any formulation to the wholesaler or retailer in respect.

Maintenance of Records

Every manufacturer and importer shall maintain all records including sales turnover of bulk drugs manufactured or imported and turnover of formulations and other records as may be directed. The Government has the power to call the records or to inspect records at the premises.

Power of Entry, Search and Seizure

Any Gazetted Officer of the Central Government or a State Government authorized by the Government for ensuring compliance with this order may enter and search place, seize any drug alongwith container, packages, etc., inspect record, seize any document such as cash memo or memo, books of accounts if found to be in contravention with provision of the order.

Any person aggrieved by any notification issued or order made may apply to the Government within fifteen days for review and the Government may make such order on the application as it may deem proper.

It is mandatory to print the words "Retail price not to exceed" on the label of the formulation of manufacturer alongwith "Local taxes extra".

Penalties

Any contravention of any provision of the order shall be punishable in accordance with the provision of the Essential Commodities Act, 1955.

Power to Exempt and Delegation of Powers

The Central Government under special conditions, may exempt any manufacturer from the operation of all or any provisions of this order after taking into consideration factors like, capital invested, number of workers employed, sales turnover, types of products manufactured, production of a new drug and indigenous and unique production.

CHAPTER 25

WTO, GATS and The Indian Patents Act, 1970 with Amendments

The enactment of the Indian Patents Act (IPA), 1970 was a revolutionary step taken by the then Government with Late Smt. Indira Gandhi as the Prime Minister of the country. This Act was aimed at giving an impetus to the industrial and economic growth within the country by adopting process-patent strategy. The effective implemtation of this Act between 1970 and establishment of World Trade Organization (WTO) in 1995 yielded desired results of increased production of different commodities by Indian industrial sector, especially in the field of drugs and pharmaceuticals. Before the enactment of IPA, 1970, the Indian pharmaceutical industry was largely depending upon import of drugs and technology, and indigenous pharmaceutical sector was not in a position to compete mighty foreign companies established long back in the country. The provision of process-patent, and not the product-patent, provided in the Act was one of the important landmarks in enhancing the bulk drug production in the country.

India's strengths of trained manpower and knowledge base, good network of research laboratories, well developed base industries, rich biodiversity, research access to vast and diverse disease populations, access to intellectual NRIs and strong agriculture base were effectively used all these years, with the legal backing of IPA, 1970 to boost production of drugs and pharmaceuticals within the country. The pharmaceutical industry in 2004 ranked fourth in the world in terms of volume (8%) of drug production. It, however, occupied thirteenth position in the world in terms of value (around 1 percent), mainly because of low cost of medicines in the country. The mandate for the Indian companies is to serve the cause of providing good quality, cheap and affordable drugs to the poor strata of the society.

The Committee constituted by the General Agreement on Tariffs and Trade (GATT) under the Chairmanship of Mr. Arthur Dunkel, the then Director-General of GATT, came out with a draft report – "Dunkel Draft" – after deliberations amongst member countries of GATT for more than five years. The draft was finally approved by participating countries, which ultimately culminated in establishment of World Trade Organization (WTO) on 1-1-1995, the international agency to act as watch-dog for implementation of the policies of TRIPS (Trade Related Intellectual Property Rights) and TRIM (Trade Related Investment Measures). The new patent regime pertaining to intellectual property has come into force from 1-1-1995 and India with some other countries, whose intellectual property legislations were not in agreement with provisions of TRIPS of WTO, were given grace period of ten years for implementation. To comply with the requirements of WTO, first amendment to the Act was effected in 1999, followed by another in 2003. Finally, the Government of India promulgated the Patent (Amendment) Ordinance which came into force w.e.f. 1st January, 2005. The Patents Rules 2003 were also modified appropriately which also came in force from 1-1-2005.

Definitions

Patent : It is a grant in the form of a document by the Government for disclosing a new invention by applicant/s.

Intellectual Property : It is a class of property emanating from human intellectual abilities and means Patents, Designs Copyrights, Geographical Indications, Trademarks, etc.

The legal rights pertaining to intellectual property are summed up as Intellectual Property Rights (IPR).

Invention : means a new product or process involving an inventive step and capable of industrial application. Inventive step means a feature that makes the invention not obvious to a person skilled in the art.

Inventions not patentable

(Under Section 3 of the Indian Patents Act, 1970.)

- Claims contrary to established natural laws.
- Mere discovery of scientific principle.

- Substance obtained by mere admixture with improved properties.
- Re-arrangement or duplication of known devices.
- Method of agriculture or horticulture.
- Mere discovery of a new property or new use of a known substance or apparatus.
- An invention, the primary or unitended use of which would be injurious to public health.
- Any process for medicinal, prophylactic, surgical, curative, diagnostic, therapeutic or other treatment of human beings.
- Plants and animals except microorganisms in whole or any part thereof including species, varieties of seed, propagation of plants and animals.
- Aggregation or duplication of known properties of traditionally known component.
- A computer program *per se* other than its technical utility to industry or a combination with hardware.
- A method or process of testing applicable during the process of manufacture.
- A literary, musical, artist, dramatic work including television and cinematographic work.
- Topography of integrated circuits.
- Presentation of information.
- A mere scheme or method or rule of performing mental act or method of playing game.
- Production, control, use or disposal of atomic energy or specific activities related to processing of enriched or radioactive substance as per Atomic Energy Act, 1962.

Types of Patents

1. **Ordinary patent :** It is granted on an ordinary application made.
2. **Patent for addition :** It is secured for an improvement or modification to granted patent. It remains in force till the original patent is valid.

3. **Patent granted under convention agreement :** It is based on an application made in a Convention Country notified under Section 133 of the Act. The convention application has to be made within twelve months from the date of the first application.

It is now possible to secure patent protection for new products and the processes of preparation in Science and Technology and also for a new hardware.

The main criteria for securing patents are :

(i) The invention must be new.

(ii) It should involve inventive step.

(iii) The invention must be industrially applicable and useful to the society.

Even though, product-patent is the key factor in post-WTO era, it does not prevent a person to get a patent for developing alternate and commercially viable new processes for the new product. The alternate and commercially viable process developed cannot be commercially used without getting the permission of the valid product patent holder.

Term of Patent

In accordance with amendment in 2002 to Section 33, the term of patent is 20 years and shall also be applicable to those patents granted under the Patents Act, 1970 from the date of amendment.

Administration

The Headquarters of Patent Office is in Kolkata with its branches in Mumbai, Delhi and Chennai. The Nodal Ministry is Ministry of Commerce and Industry, Department of Industrial Policy and Promotion, Government of India, Udyog Bhavan, New Delhi.

Procedure for Patenting

The application filed in the prescribed forms along with the prescribed fees accompanied by provisional is required to be made by inventor or his legal representative to the officer-in-charge of patent office. The provisional specifications describe essential features of invention, whereas complete specifications include full description, statement, detailed method, scope, utility, etc. The complete specifications should be produced within twelve months of filing of provisional specification. No extension beyond this period is possible in case of applications filed on or after 1-1-2005.

The Patents Office refers the contents of complete specification to scientific adviser or examiner for scrutiny. Examiner's report is obtained within eighteen months either to accept or modify or reject. The new amendment introduced is the facility for withdrawing the application before publication on the expiry of eighteen months from the date of filing. The Controller General of Patents, in case of acceptance, notifies and advertizes in Official Gazette. Any objection for publication of patent from public should be received within one year from the date of publication of grant of patent (Notice of Opposition). The opportunity is given to hear and interact and only on satisfying all conditions, the patent is granted with seal of office and entry of date of sealing in register.

Requirements

The documents required in duplicate for filing an application for patent in India are as follows :

1. Application **Form 1** suitably modified.
2. Provisional Certification **(Form 2)** or Complete Specification **(Form 2)**.
3. Abstract of invention.
4. Any drawings, if necessary.
5. Statement and undertaking **(Form 3)**.
6. Prescribed fees.

Inventions Patentable

Development and process for the preparation of new compounds, new materials, new formulations or synergistic compositions; improvement of an existing process for production of known compounds, known compositions and known materials; and development of new medical kit, new machine/device, new hardware incorporating a software and improved machine device are patentable.

Patent Search

The patents are categorized under the National Patent Systems of every country as well as under an International Patent Classification System (IPC). The IPC system of classification categorizes the entire Science and Technology into eight main headings, each heading is further classified into about 95000 sub-headings.

Rights of Patentee

The rights of the patentee are defined in Section 48 of the Act. The grant of a product patent prevents a person in commercially utilizing the process he has developed for the preparation of the product patent holder. The patent holders for respective invention can come together and enter into joint agreement for mutual benefits, if the alternate process developed has significant commercial utility.

The patentee has an exclusive right to make use of the patent within the period of patent, sell or distribute such an article. He has right to prevent third parties who do not have his consent from the act of using that process.

For Government use under special provisions there is no restriction for importing or making use of provision. The patent is required to be surrendered if complaint is made and upheld by the order of High Court.

Infringement of Patent

When the legal rights in a valid patent are commercially exploited without the permission of the patentee, it amounts to infringement of patent.

The relief, which a court may grant in any suit for infringement, includes an injunction and at the option of the plaintiff either damages or an account of profits.

Other Features

- The legal rights in intellectual property secured (except for copyright) in one country cannot be enforced in another country.
- For availing the benefits of the Patent Cooperation Treaty (PCT) Rules, a single application (International Application) has been devised in respect of designated member countries of PCT, where protection is desired. Afterwards, the applicant, however, has to make corresponding applications in the member countries of PCT.
- The amendments effected have a provision for establishment of an "Appeal Board", the qualifications for the members of the Board which shall look into IP disputes and provide speedy, timely and inexpensive disposal of appeals.

- "Protection of National Security" clause has been inserted, whereby the Central Government can take any action including revocation of a patent in the interest of the security of India by issuing notification in the Official Gazette to that effect.

The Patent Amendment 2005 covering product protection has provided an opportunity for domestic players to invest in R & D and bring out successful drugs at a greatly reduced cost.

General Agreement on Trade in Services (GATS)

The Indian delegation at WTO has submitted revised offer in August 2005 in the ongoing negotiations under GATS.

The offers were initially made in the Doha Round in sectors/sub-sectors covered in the commitments made in Uruguay Round. The seven sectors of services covered were as follows:

1. Health related and social services
2. Tourism and travel related services
3. Communication services
4. Financial services
5. Business services
6. Construction and related engineering services
7. Transport services.

The revised offer includes four other sectors which were not covered earlier. These services are:

1. Distribution services
2. Education services
3. Environmental services
4. Cultural, sporting and recreational services.

GATS provides for four modes of supply of services.

Mode 1 – Cross-border supply

Mode 2 – Consumption abroad

Mode 3 – Commercial presence

Mode 4 – Presence/movement of natural persons.

India and many other developing countries are in Modes 1 and 4. Mode 4 interest is due to the presence of a large skilled and competitive workforce. It will enable India to take advantage of Mode 1 due to core competence in Information Technology related services.

India had already made a substantial Mode 4 initial offer by including all categories of natural persons including independent professionals.

The sector of Business Services covers medical and dental services, services provided by nurses, midwives, pharmacists, physiotherapists and other paramedical services are included. Section 3 offered includes higher education services.

India spends only 0.9 percent of its GDP on health care in public sector forcing the majority of the people to turn to private health systems that are often beyond their reach. The Indian Government has recognized health as an inalienable human right that every individual can claim. So long as wide health inequalities exist in our country and access to essential health care is not universally assured, it is feared that we would fall short in both economic planning and in our moral obligations to all citizens.

The developments in post-WTO era related to new patent regime and GATS should be accepted as challenges and converted into opportunities to further strengthen our infrastructure of pharmaceutical industries and their R & D set-up and also global exploitation and utilization of our professional services and skills.

Chapter 26

Good Laboratory Practice (GLP)

Good Laboratory Practice (GLP) is defined, in accordance with OECD Principles, as "a quality system related to the organizational process and the conditions under which non-clinical health and environmental safety studies are planned, performed, monitored, recorded, archived and reported." The basic principle of GLP is to promote the development of quality test data and also to provide a tool to ensure a sound approach to the management of laboratory studies, including conduct, reporting and archiving. It provides a set of standards for the quality, reliability and integrity of studies, reporting of verifiable conclusions and traceability of data. The institutions are required to define the roles and assign responsibilities to the staff, in order to ensure good operational management of each study and to focus on those aspects of study, execution (planning, monitoring, reporting, recording, and archiving) that are of significance for the reconstruction of the whole study. It is not permissible to partially implement GLP requirements and still claim GLP compliance. GLP compliance can be claimed for any facility provided full array of the GLP rules are made applicable.

In relation to pharmaceutical development, the GLP Principles apply to the studies of following types :

- Non-clinical, i.e. mostly studies on animals or *in -vitro*, including analytical aspects.
- Studies designed to obtain data on the properties and/or the safety of items in relation to human health and/or the environment.

- Studies that are intended to be submitted to a national registration authority with an objective of registering or licensing the tested substance or any product derived therefrom.
- GLP requirements for non-clinical laboratory studies which are conducted to evaluate drug safety include –
  1. Single dose toxicity
  2. Repeated dose toxicity (sub-acute and chronic)
  3. Reproductive toxicity (fertility, embryo-foetal toxicity, teratogenicity, peri-/post-natal toxicity)
  4. Carcinogenic potential
  5. Mutagenic potential
  6. Pharmacodynamic studies designed to test the potential for adverse effects (safety pharmacology)
  7. Toxicokinetics
  8. Local tolerance studies, including phototoxicity, irritation and sensitisation studies, or testing for suspected addictive and/or withdrawal effects of drugs.

GLP is not directly concerned with the scientific design of studies. The scientific design may be based on test guidelines and its scientific value is judged by the Drug Regulatory Authority that provides marketing authorisation. Adherence to GLP removes many sources of error and uncertainty, contributing to the overall credibility of the study.

GLP Principles apply to studies planned and conducted in a manufacturer's laboratory, at a contract or subcontract facility, or in a university or public sector laboratory. They are independent of the site where studies are performed.

Many sources of systematic error and artefacts may be avoided by application of technically valid and approved Standard Operating Procedures (SOPs).The GLP Principles shall indirectly optimise the scientific yield of studies.

While implementing GLP in a test facility, and particularly during training, it is necessary to clearly differentiate between the formal, regulatory use of the term Good Laboratory Practice and the general application of "good practices" in scientific investigations. The term "Good Laboratory Practice" is not a trade-mark protected term. Therefore, it is likely that any laboratory may consider that it is following good practices in its daily work without GLP compliance. **It must be clearly understood that only adherence to, and compliance with, all the requirements of the OECD-GLP Principles constitutes real compliance with GLP. Hence the use of identical terminology to describe quality practices/parameters beyond the scope of GLP is misleading, and such a practice should be strongly discouraged.**

The GLP Principles set out the requirements for the appropriate management of non-clinical safety studies which helps the researcher to perform studies in compliance with his/her own pre-established scientific design. The GLP regulations are not concerned with the scientific or technical content of the studies *per se*. The regulations do not aim to evaluate the scientific value of the studies. This task is first for senior scientists working on the research programme, then for the registration authorities, and eventually for the international scientific community as a whole. The GLP requirements for proper planning, controlled performance of techniques, appropriate monitoring of activities, systematic recording of all observations, and for complete archiving of all raw data generated are necessary to ensure many sources of error. GLP emphasizes the significance of the following.

1. Resources: organisation, personnel, facilities and equipment,
2. Rules: protocols and standard operating procedures (SOPs),
3. Characterisation: test items and test systems,
4. Quality assurance: independent monitoring of research processes,
5. Results: raw data, final report and archives.

CHAPTER 27

Adverse Drug Reactions (ADRs)

No medicine is totally safe for everyone in all circumstances. Evidences of Adverse Drug Reactions (ADRs) continue to pour in from different parts of the globe. The hazard may be sometimes insignificant and acceptable in relation to the therapeutic utility of drugs. All hazards may not be known before a drug is marketed. It is in this context, study of pharmacovigilance acquires special significance.

Adverse Drug Reactions (ADRs), according to the International Conference on Harmonisation of Technical Requirements for Registration of Pharmaceuticals for Human Use, is defined as a response to a drug which is noxious and unintended and which occurs at doses normally used in human beings for diagnosis, prophylaxis or therapy or for modification of physiological function. ADR covers an exaggerated drug response, an unwanted influence on an organ system different from that being treated and an allergic or hypersensitivity reaction. It also includes a drug interaction that causes an increased or diminished response.

Pharmacovigilance is the systematic study of drug-related injuries for the purpose of making warning or withdrawal recommendations for pharmaceutical products. It targets the monitoring of safety of drugs and aims at promoting patient-care and patient safety in relation to the use of medicines. Pharmacovigilance helps to assess the risk-benefit profile of medicines and thus, encouraging safer and more effective use of medicines.

With the discovery and development of new drugs and their formulations and their marketing in India, a need for vibrant pharmacovigilance system in the country was acutely felt to facilitate the strategy for protection of the population from the potential harm that may be caused by some of these new drugs. With the new patent regime coming in force from 2005, it was widely believed that India would become hub for new drug clinical trials. In view of vastness of our country with multi-ethnic population, it was felt that there is an urgent requirement to have a viable Adverse Drug Reaction (ADR) data capturing mechanism.

National Pharmacovigilance Advisory Committee (NPAC)

The Indian Central Drug Regulatory Authority has proposed to set up a pharmacovigilance programme in the country with following objects.

1. To improve public health and safety in relation to use of medicine.
2. To contribute to regulatory assessment of benefit, harm, effectiveness and risks of medicines.
3. To strengthen pharmacovigilance through proper understanding, education and clinical training, and
4. To improve patient care and safety in relation to use of medicines and different medical and paramedical interventions.

The National Pharmacovigilance Advisory Committee (NPAC) oversees the performance of different zonal, regional and peripheral Pharmacovigilance Centres. At present, there are twenty four Peripheral Centres for recording Adverse Events (AE), six Regional Centres to scrutinize the data received from Peripheral Centres, and two zonal centres for causality analysis of reported AEs and for providing training and general support to Regional Centres. The composition of National Pharmacovigilance Advisory Committee is represented by following members.

(i) Director General of Health Services – Chairman

(ii) Drugs Controller General of India – Member Secretary
MEMBERS

(iii) Director General, ICMR

(iv) Director, Drugs Division, Ministry of Health & Family Welfare

(v) Zonal Centre Coordinator from AIIMS

(vi) Zonal Centre Coordinator from SGC Medical College, Mumbai
(vii) Expert in Pharmacology
(viii) Expert in General Medicine
(ix) Expert in Clinical Pharmacology
(x) Expert in Forensic Medicines
(xi) Expert in Pathology
(xii) Expert in Toxicology
(xiii) Expert in Epidemiology
(xiv) Expert in Pharmacy
(xv) Expert in Drug Information
(xvi) Expert in Phytotherapy

All members are nominated by the Ministry of Health & Family Welfare. The Centre is located at CDSCO (HQ), Delhi.

The National Pharmacovigilance Centre, AIIMS, New Delhi invites reports of all suspected adverse reactions to drugs including herbal, traditional and of alternative remedies. The information pertaining to adverse drug reactions of all types suspected of causing hospitalization, prolongation of hospitalization, danger to life, death, birth defects, increased cost of treatment and absence from productive activity is collected by the National Pharmacovigilance Centre. It is then processed on scientific lines and placed before the (NPAC) for taking remedial measures in the interest of the society at large.

Chapter 28

List of Prohibited Drugs in India

The drugs that are prohibited for manufacture and sale in India under section 26A of the Drugs and Cosmetics Act, 1940 are listed below.

1. Prohibited Drugs

Amidopyrine, phenacetin, nialamide, practolol, methapyrilene and its salts, methaqualone, chloral hydrate as a drug, fenfluramine and dexfenfluramine.

2. Prohibited Fixed Dose Combinations

(i) Vitamins with anti-inflammatory agents and tranquilizers
(ii) Atropine with analgesics and antipyretics
(iii) Strychnine and caffeine in tonics
(iv) Yohimbine and strychnine with testosterone and vitamins
(v) Iron with strychnine, arsenic and yohimbine
(vi) Sodium bromide / chloral hydrate with other drugs
(vii) Antihistaminic with anti-diarrhoeals
(viii) Penicillin with sulphonamides
(ix) Chloramphenicol with any other drug for internal use
(x) Vitamins with anti-TB drugs except, combination of isoniazid with vitamin B6
(xi) Crude ergot preparations except, those containing ergotamine, caffeine, antihistamines and analgesics
(xii) Corticosteroids with any other drug for internal use

(xiii) Tetracycline with Vitamin C

(xiv) Vitamins with analgesics

(xv) Hydroxyquinoline group with any other drug except, for external use preparations

(xvi) Anabolic steroids with other drugs

(xvii) Sedatives / hypnotics / anxiolytics with analgesics-antipyretics

(xviii) Histamine H-2 receptor antagonists with antacids except, those approved by DCGI

(xix) Essential oils with alcohol having more than 20% proof except, preparations given in I.P

(xx) Ethambutol with INH other than INH-ethambutol (200 mg – 600 mg, and 300 mg – 800 mg)

(xxi) More than one antihistamine.

(xxii) Any anthelmintic except, piperazine/santonin with cathartic or purgative

(xxiii) Salbutamol or any other bronchodilator with centrally acting anti-tussive and/or antihistamine

(xxiv) Laxatives and/or antispasmodic drugs in enzyme preparations

(xxv) Metoclopramide with systemically absorbed drugs except, fixed dose combination of metoclopramide with aspirin / paracetamol

(xxvi) Antitussive with antihistamine having high atropine-like activity in expectorants

(xxvii) Analgin with any other drug

(xxviii) Antidiarrhoeals with electrolytes

(xxix) Oxyphenbutazone or phenylbutazone with any other drug

(xxx) Pectin and/or Kaolin with any drug which is systemically absorbed

(xxxi) Dextropropoxyphene with any other drug other than anti-spasmodics and/or non-steroidal anti-inflammatory drugs

(xxxii) Diazepam with diphenhydramine hydrochloride

(xxxiii) A drug prescribed for standards in Second Schedule to the Act with an Ayurvedic, Siddha or Unani drugs

(xxxiv) Oestrogen and progestin (other than oral contraceptive) containing per tablet estrozen content of more than 50 mcg and progestin content of more than 3 mg

(xxxv) Injectable preparations containing synthetic oestrogen and progesterone

(xxxvi) Combination of rifampicin, isoniazide and pyrazinamide except those which provide daily adult dose of rifampicin (450 mg-600 mg), isoniazid (300mg-400mg) and pyrazinamide (1000 mg-1500 mg)

(xxxvii) Histamine H-2 receptor antagonists with antacids except those approved by DCGI

(xxxviii) Patent and proprietary medicines of fixed dose combinations of essential oils with alcohol more than 20% proof except, preparations given in the Indian Pharmacopoeia

Prohibited Other Formulations

(i) Mepacrine hydrochloride in any dosage form for use for female sterilization or contraception

(ii) Dovers Powder IP. & its tablets

(iii) Antidiarrhoeal formulations containing phthalyl sulphathiazole or sulphaguanidine or succinyl sulphathiazole

(iv) Antidiarrhoeal formulations containing kaolin or pectin or activated charcoal or attapulgite

(v) Antidiarrhoeal formulations containing neomycin or streptomycin or dihydrostreptomycin or their salts or esters

(vi) Pencillin skin/eye ointment

(vii) Tetracycline liquid oral preparations

(viii) Oxytetracycline liquid oral preparations

(ix) Demeclocycline liquid oral preparations

(x) All preparations containing chloroform exceeding 0.5% w/w or v/v

(xi) Preparations for curing cough associated with asthma containing centrally acting antitussive and/or antihistamine

(xii) Liquid oral tonic preparations containing glycerophosphates and/or other phosphates and/or C.N.S stimulant and preparations containing alcohol more than 20% proof

(xiii) Liquid oral antidiarrhoeals or any other dosage form for pediatric use containing diphenoxylate lorloperamide or atropine or belladona including their salts or esters or hyoscyamine or their extracts

(xiv) Liquid oral antidiarrhoeals or any other dosage form for pediatric use containing halogenated hydroxyquinolines

ANNEXURE

Important Professional Associations Related to Pharmacy

Several professional associations in the country are engaged in wide spectrum of activities related to the pharmacy. These associations provide a suitable platform for professional deliberations which are aimed at raising the standards of the profession. Important associations are listed below.

S.No.	Name	Address
1.	Indian Pharmaceutical Association (IPA)	Kalina, Santacruz East Mumbai – 400 098 Ph: 022-26671072 Email: ipacentre@ipapharma.org
2.	Association of Pharmaceutical Teachers of India (APTI)	Al-Ameen College of Pharmacy, Lalbagh Main Gate, Hosur Road, Bangalore – 560 027 Ph: 080 - 22234619 Email: aptialerts@yahoo.com
3.	Indian Hospital Pharmacists Association (IHPA)	Dept of Pharmacy, St.stephen's Hospital, Tis Hazari, Delhi – 110 054 Ph: 011-23966021 Email: bectrop@gmail.com

Contd....

S.No.	Name	Address
4.	Indian Pharmacy Graduates Association (IPGA)	E2, Shopping Complex, A Block, Meera Bagh, New Delhi – 110 087 Ph: 011-45612324 Email: info@ipga.in
5.	All India Drug Control Officers Confederation (AIDCOC)	28, Rajiv, Bandra-Kurla Complex, Mumbai – 400 051 Ph: 022-26435370 Email: aidcocsd@vsnl.com
6.	Indian Pharmaceutical Congress Association (IPCA)	Annual Conference every year with the participation of IPA, APTI, IHPA,IPGA & AIDCOC Presently, the head quarters is at Chennai. The IPC Conference is held at different places in the country.
7.	Indian Drug Manufacturers Association (IDMA)	102-B, Poonam Chambers, Dr.A.B.Road, Worli, Mumbai – 400 018 Ph: 022-4974308 Email: idma@vsnl.com
8.	Pharmaceutical Export Promotion Council (PHARMEXIL)	101, Aditya Trade Centre, Ameerpet, Hyderabad – 500 038 Ph: 040-55458488 Email: info@pharmexcil.com
9.	Organization of Pharmaceutical Producers of India (OPPI)	Cooks' Building, Dadabhay Naoroji Road, Mumbai – 400 001 Ph: 022-2045509 Email: indiaoppi@vsnl.com
10.	All India Small Scale Pharmaceutical Manufacturers Association (AISSPMA)	26, Bhargav Lane Nityanand Marg, Civil Line Delhi – 110 054 Ph: 011-22919140

Contd....

S.No.	Name	Address
11.	All India Organization of Chemists and Druggists (AIOCD)	201,Safalya Bldg, IInd Floor, Opp Jaigopal Industries Estate, Baburao Parulekar, Dadar West, Mumbai – 400 028 Ph: 022-24306889 Email: aiocd@vsnl.net
12.	Bulk Drug Manufacturers Association (BDMA)	A-24, View Towers, Lakadi Ka Pul, Hyderabad – 500 004 Ph: 040-3322142 Email: info@bdm.assn.org
13.	Pharmaceutical and Allied Manufacturers and Dealers Association (PAMDAL)	C/o Bombay Chambers of Commerce & Industry Mackinnon Mackenzie Building , Ballard Estate, Mumbai – 400 030 Ph: 022-2614681 Email: pamdal@bombaychamber.com
14.	Confederation of Indian Industry (CII)	23, Industrial Area, Lodi Estate, New Delhi – 110 003 Ph: 011-24621874 Email: td@co.cii.ernet.in
15.	Federation of Indian Chambers of Commerce and Industry (FICCI)	Federation House, Tansen Marg, New Delhi – 110 001 Ph: 011-23738760 Email: ficci.bisnet@gems.vsnl.net.in
16.	Basic Chemicals, Pharmaceutical & Cosmetics Export Promotion Council (CHEMEXCIL)	Jhansi Castle, 7, Cooperage, Mumbai – 400 039 Ph: 022-22021288 Email: chemexcil@vsnl.com

References

Bulletins and Minutes of the Meetings of Central Pharmacy Council of India, Pharmacy Council of India, Kotla Road, Delhi, 1994-2003.

Drugs and Cosmetics Act and Rules, 3rd Edition, by S.W. Deshpande and Nilesh Gandhi, Susmit Publishers, Mumbai, 2004.

Drugs and Cosmetics Act, 1940 by Vijay Malik, Eastern Book Company, Lucknow, 2003.

Good pharmacy practice Guidelines, Indian Pharmaceutical Association, Mumbai, 2002.

Hand Book of Drug Laws, 10th Edition by M.L. Mehra, Universal Law Publishing Company, Delhi, 2005.

History of Pharmacy in India and Related Aspects – Pharmacy Practice, Harkishan Singh, Volume 3, Vallabh Prakashan, Delhi, 2002.

History of Pharmacy in India and Related Aspects, Volume 2 : Pharmaceutical Education, Harkishan Singh, Vallabh Prakashan, Delhi.

Laws of Drugs by M.P. George, Magna International Publications, Kollam, 1999.

Model Public Health Act, Ministry of Health and Family Welfare, Government of India, New Delhi, 1987.

Modifications in Drug Policy, 1986, Ministry of Chemicals and Fertilizers, Government of India, 1994.

National Blood Policy, National AIDS control Organisation (NACO), Ministry of Health and Family Welfare, Government of India.

Law Relating to Drugs and Cosmetics, Vijay Malik. 22nd Edition, Eastern Book Company, Lucknow, 2011.

Schedule - y (Amended Version-2005), Academy for Clinincal Excellence, Kalina, Mumbai.

Schedule M (India) & WHO TRS 908, GMP, Updated 2008 Edition

National Policy on Education - 1986, Ministry of Human Resource Development, Government of India, 1986.

Patents by N.R. Subbaram, 2nd Edition, Pharma Book Syndicate, Hyderabad.

Quintessential Wisdom I and II. Presidential Addresses of All India Pharmaceutical Conferences, edited by M.C. Bazaz and D.B.A. Narayana, New Delhi, 1998 and 2000.

Report of the Drugs Enquiry Committee, Government of India, Central Publication Branch, Calcutta, 1931.

The All India Council for Technical Education Act, 1987, AICTE, New Delhi, 1987.

Indian Pharma Reference Guide, 2007.08, Kongposh Publication Pvt. Ltd., New Delhi.

A Text Book of Pharmacy Practice by K.G.. RaviKumar and B.D. Miglani, Career Publications, Nashik, 2009.

The Drugs and Cosmetics Act, 1980, Government of India, Ministry of Health and Family Welfare, Delhi.

The Medical and Toilet Preparations (Excise Duties) Act, 1955, Delhi Law House, Delhi, 1989.

The Pharmacy Act, 1948, Ministry of Health and Family Welfare, Government of India, New Delhi.

The University Grants Commission Act, 1956, U.G,.C., Bahadur Shah Zafar Marg, New Delhi, 2000.

Hnadbook for Community Pharmacists, authored by Dr. Astmaram Pawar, Career Publications, Nashik, 2007.

'Editing Pharmacy' by B.D. Miglani, APTI, Bangalore, 2011

Indian Pharmacopoeia Ministry of Health & Family Welfare, Govt. of India, N. Delhi, 1996.

Report of Expert Committee on Medical Stores Organisation, Govt. of India, N. Delhi, 1997.

All Acts related to Pharmacy Govt. of India Publications.

Index